415291   £3.00

THE OPEN UNIVERSITY

Mathematics: A Third Level Course

Complex Analysis   Units 10 and 11

The Calculus of Residues
Analytic Functions

Prepared by the Course Team

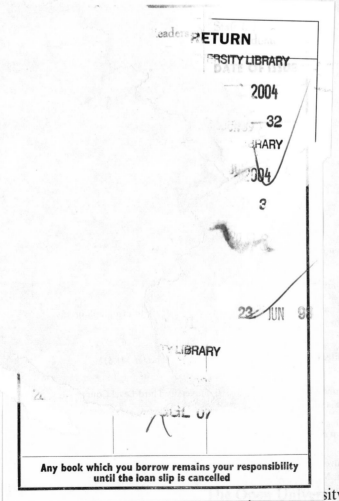
sity Press

The Open University Press, Walton Hall, Milton Keynes.

First published 1975

Produced in Great Britain by
Technical Filmsetters Europe Limited, 76 Great Bridgewater Street, Manchester M1 5JY.
ISBN 0 335 05553 2

This text forms part of the correspondence element of an Open University Third Level Course. The complete list of units in the course is given at the end of this text.

For general availability of supporting material referred to in this text, please write to the Director of Marketing. The Open University, P.O. Box 81, Milton Keynes, MK7 6AT.

Further information on Open University courses may be obtained from The Admissions Office, The Open University, P.O. Box 48, Milton Keynes, MK7 6AB.

1.1

CONTENTS

Unit 10 The Calculus of Residues

Conventions

Before working through this text make sure you have read *A Guide to the Course: Complex Analysis.*

References to units of other Open University courses in mathematics take the form:

Unit M100 13, Integration II.

The set book for the course M231, Analysis, is M. Spivak, *Calculus*, paperback edition (W. A. Benjamin/Addison-Wesley, 1973). This is referred to as:

Spivak.

Optional Material

This course has been designed so that it is possible to make minor changes to the content in the light of experience. You should therefore consult the supplementary material to discover which sections of this text are not part of the course in the current academic year.

10.0 INTRODUCTION

Up to this point in the course, our study of the integration of functions of a complex variable has led us in two different directions. On the one hand, we have been concerned with the *theory* of the integration of functions along contours, and on the other, we have been developing *techniques* which can be used when we actually want to evaluate a particular integral. Let us look at each of these in turn.

As far as the theory is concerned, we introduced the idea of integration along a contour in *Unit 4, Integration,* and we saw in *Unit 5, Cauchy's Theorem I,* that the integral of an analytic function along a suitable closed contour is always zero. This is the simple version of Cauchy's Theorem, since it applies only to a rather restrictive class of regions, the star regions. It was not until *Unit 9, Cauchy's Theorem II,* that we introduced the general version, which shows that the conclusions of Cauchy's Theorem hold for any simply-connected region— a great improvement. Once we had this result, we were able to extend Cauchy's Formula to general closed contours, and finally to prove the Residue Theorem:

The Residue Theorem

Let R be a simply-connected region, $\alpha_1, \ldots, \alpha_n$ be points of R, and f be a function analytic on R except for singularities at $\alpha_1, \ldots, \alpha_n$ (so that f is analytic on $R - \{\alpha_1, \ldots, \alpha_n\}$. Let Γ be any simple-closed contour in R not passing through any of $\alpha_1, \ldots, \alpha_n$. Then

$$\int_\Gamma f(z)dz = 2\pi i \sum_{j \in S} \text{Res}(f, \alpha_j),$$

where $S = \{j : 1 \leqslant j \leqslant n$ and α_j lies inside $\Gamma\}$. In other words,

$$\int_\Gamma f(z)dz = 2\pi i \times \text{ the sum of the residues of } f \text{ at those singularities of } f \text{ lying inside } \Gamma.$$

So the Residue Theorem can be regarded as the last in a chain of *theoretical* results on complex integration, a fact which should have emerged very clearly from *Unit 9.*

In this unit, we shall take a different point of view. Our aim is to show that much of the importance of the Residue Theorem derives from its applications— especially to the evaluation of integrals and the summation of series—in addition to any theoretical interest it may have.

To see how the Residue Theorem fits in, let us look back and see what methods we have used up to now for calculating contour integrals. You will remember that in *Unit 4,* the calculation of integrals was first effected by choosing a convenient parametrization for the contour, and just 'plugging it in'. Then we introduced the Fundamental Theorem of Calculus, for integrating the derivatives of known functions, and later Cauchy's Formulas, for evaluating integrals of the form

$$\int_\Gamma \frac{f(z)}{(z - \alpha)^n} dz, \quad \text{where } n \text{ is a positive integer.}$$

We also introduced such techniques as integration by parts and substitution.

The next step forward occurred at the end of *Unit 8, Singularities,* in which we introduced Laurent series. If f is analytic on some punctured disc with centre α then f can be expressed, in a unique way, as a Laurent series of the form

$$f(z) = \cdots + \frac{a_{-2}}{(z - \alpha)^2} + \frac{a_{-1}}{z - \alpha} + a_0 + a_1(z - \alpha) + a_2(z - \alpha)^2 + \cdots,$$

convergent on that punctured disc. The coefficients a_n are given by the formula

$$a_n = \frac{1}{2\pi i} \int_\Gamma \frac{f(z)}{(z - \alpha)^{n+1}} dz, \quad \text{where } \Gamma \text{ is any circle centre } \alpha \text{ lying in the}$$

disc.

7

The coefficient a_{-1}, called the *residue* of f at α, Res(f, α), is given by the formula

$$\text{Res}(f, \alpha) = a_{-1} = \frac{1}{2\pi i} \int_{\Gamma} f(z)dz.$$

At the end of *Unit 8*, the whole argument was turned around, and the value of the residue was used to help us calculate the value of the integral $\int_{\Gamma} f(z)\,dz$. In fact, *Unit 8* concluded with a substantial number of integrals of the form $\int_{\Gamma} f(z)\,dz$ (where Γ was a circle, and f had a pole inside Γ), which could be evaluated by calculating the relevant residue and multiplying by $2\pi i$.

In this unit, we shall carry these ideas a lot further. One of our aims is to calculate a large number of integrals by combining the technique just described with an application of the Residue Theorem. The basic idea is to try to evaluate an integral of the form $\int_{\Gamma} f(z)\,dz$ by calculating the residues of f at the poles of f inside Γ, and deducing the result from the Residue Theorem. We shall give you a lot of practice in using this sort of argument, and you will also see some of the applications to which it can be put. In fact, all of the integrals we evaluate *can* be evaluated by other means, but the residue approach is usually far quicker and easier to use, as well as being a general technique which is applicable in a large number of cases.

We start this unit, in Section 10.1, with a section on the calculation of residues, in which you will have an opportunity to review some of the material presented at the end of *Unit 8*, as well as to learn some new techniques for calculating residues. This is followed, in Section 10.2, with a description of the way residues can be used to evaluate certain types of contour integral, and also, surprisingly, to evaluate some real trigonometric integrals involving cos and sin, such as

$$\int_0^{2\pi} \frac{d\theta}{2 + \sin \theta} \quad \text{and} \quad \int_0^{2\pi} \frac{\sin \theta}{\cos^2 \theta - \sin \theta} d\theta.$$

In Section 10.4, the most important section in this unit, we shall show you how residues can be used to evaluate certain real improper integrals of the form

$$\int_{-\infty}^{\infty} f(t) \cos \beta t \, dt \quad \text{and} \quad \int_{-\infty}^{\infty} f(t) \sin \beta t \, dt,$$

where f is a rational function. Some of the ideas developed in this section are then used in Section 10.6 to sum various series, such as

$$\sum_{n=1}^{\infty} \frac{1}{n^4} \quad \text{and} \quad \sum_{n=1}^{\infty} \frac{1}{a^2 - n^2},$$

by calculating the residues of some seemingly unrelated functions. Finally, in Section 10.8, we shall use the Residue Theorem to obtain information about the number of zeros and poles of a given function inside a given contour.

As you can see from the previous paragraph, our aim in this unit is to show you some of the wide variety of uses to which the Residue Theorem can be put, and to give you a lot of practice in using it. Because of this, you will probably find this unit considerably more 'techniques-orientated' than previous units, and you will need to spend a larger than usual proportion of your time working your way through problems.

Television and Radio

In the fifth radio programme associated with the course we shall look at several of the techniques for calculating residues, and use them to evaluate some contour integrals. In the seventh television programme we shall extend this technique to evaluate certain real integrals.

10.1 CALCULATING RESIDUES

In this section we shall look at various methods we can use when calculating residues. Some of these you will have met in *Unit 8*; others will be new. As you will see, there are several self-assessment questions throughout this section. You should work through all of these, and you should not proceed until you are satisfied you can do them. Many of the residues you calculate in this section will be needed later in the unit.

Using the Laurent Series

Since the residue of a function f with a singularity at a point α is simply the coefficient a_{-1} of $(z - \alpha)^{-1}$ in the Laurent series

$$f(z) = \cdots + \frac{a_{-2}}{(z - \alpha)^2} + \frac{a_{-1}}{z - \alpha} + a_0 + a_1(z - \alpha) + a_2(z - \alpha)^2 + \cdots,$$

we can find the residue of f at α by looking at the relevant part of this series. The following examples will show you some of the ways in which this can be done.

Example 1

Find the residue of $z \longrightarrow \dfrac{1}{z} + z$ at 0.

Solution

The Laurent series is clearly

$$\cdots + \frac{0}{z^3} + \frac{0}{z^2} + \frac{1}{z} + 0 + z + 0z^2 + 0z^3 + \cdots,$$

and so the residue (the coefficient of z^{-1}) is 1.

Example 2

Find the residue of $z \longrightarrow \dfrac{\sin z}{z^4}$ at 0.

Solution

The Taylor series for sin at 0 is

$$\sin z = z - \frac{z^3}{3!} + \frac{z^5}{5!} - \cdots,$$

and so, dividing by z^4, we get

$$\frac{\sin z}{z^4} = \frac{1}{z^3} - \frac{1}{6z} + \frac{z}{5!} - \cdots.$$

The residue is therefore $-\frac{1}{6}$.

Sometimes a little manipulation is needed, in order to get the series into the right form.

Example 3

Find the residue of $z \longrightarrow \dfrac{1}{z(z - 1)^2}$ at 1.

9

First Solution

We want to expand the given function in powers of $z - 1$. But

$$\frac{1}{z} = \frac{1}{1 + (z - 1)}$$

$$= 1 - (z - 1) + (z - 1)^2 - (z - 1)^3 + \cdots, \quad \text{if } |z - 1| < 1,$$

so that

$$\frac{1}{z(z - 1)^2} = \frac{1}{(z - 1)^2}[1 - (z - 1) + (z - 1)^2 - (z - 1)^3 + \cdots]$$

$$= \frac{1}{(z - 1)^2} - \frac{1}{z - 1} + 1 - (z - 1) + \cdots.$$

The residue is therefore -1.

We can sometimes simplify residue calculations by making a simple substitution. Since the residue of f at α is the coefficient of $(z - \alpha)^{-1}$ in the Laurent series for f about α, it follows, on substituting $z = \alpha + h$, that the residue is simply the coefficient of h^{-1} in the resulting Laurent series about 0. A couple of examples will make this clear.

Example 3 (*again*)

Find the residue of $z \longrightarrow \dfrac{1}{z(z - 1)^2}$ at 1.

Second Solution

Let $z = 1 + h$. Then

$$\frac{1}{z(z - 1)^2} = \frac{1}{h^2}(1 + h)^{-1} = \frac{1}{h^2}(1 - h + h^2 - \cdots)$$

and the residue—the coefficient of h^{-1}—is -1, as before.

Example 4

Find the residue of $z \longrightarrow \dfrac{\pi}{\sin \pi z}$ at n, where n is an integer.

First Solution

Let $z = n + h$. Then

$$\frac{\pi}{\sin \pi z} = \frac{\pi}{\sin \pi(n + h)} = \frac{\pi}{\sin n\pi \cos \pi h + \cos n\pi \sin \pi h} = \frac{(-1)^n \pi}{\sin \pi h}.$$

To find the coefficient of h^{-1} in this expression, we write

$$\frac{1}{\sin \pi h} = \left(\pi h - \frac{(\pi h)^3}{3!} + \cdots\right)^{-1}$$

$$= \frac{1}{\pi h}\left(1 - \left[\frac{(\pi h)^2}{6} - \cdots\right]\right)^{-1}$$

$$= \frac{1}{\pi h}\left(1 + \left[\frac{(\pi h)^2}{6} - \cdots\right] + \left[\frac{(\pi h)^2}{6} - \cdots\right]^2 + \cdots\right),$$

by the Binomial Theorem,

$$= \frac{1}{\pi h}\left(1 + \frac{(\pi h)^2}{6} + \cdots\right).$$

It follows that the required residue is $(-1)^n \pi \cdot \frac{1}{\pi} = (-1)^n$, the coefficient of h^{-1}.

(This result is important, and will be used in Section 10.6. A simpler way of deriving it will be given later in this section.)

Now for some exercises for you to try; all of them can be done by using the relevant Laurent series.

Self-Assessment Questions

1. Find the residues of the following functions at 0.

    (i) $z \longrightarrow \dfrac{1}{z^2}$;

    (ii) $z \longrightarrow \dfrac{1}{(z-1)^2}$;

    (iii) $z \longrightarrow \dfrac{e^z}{z^n}$, where n is a positive integer;

    (iv) $z \longrightarrow \dfrac{\sin z}{z}$;

    (v) $z \longrightarrow z^2 \sin \dfrac{1}{z}$.

2. Find the residues of the following functions at the points α indicated.

    (i) $z \longrightarrow \dfrac{1}{z^2 + 1}$, $\qquad \alpha = i$;

    (ii) $z \longrightarrow \dfrac{ze^{iz}}{(z - \pi)^2}$, $\qquad \alpha = \pi$;

    (iii) $z \longrightarrow \dfrac{1}{z^2 \sin z}$, $\qquad \alpha = 0$;

    (iv) $z \longrightarrow \dfrac{(z^2 + 1)^n}{z^{n+1}}$ where n is a positive integer, $\qquad \alpha = 0$.

3. Is the following statement true or is it false?

    If f has residue ω at the point α, then f^2 has residue ω^2 at the same point.

    If you think it is true, give a proof; otherwise, give a counter-example.

Solutions

1. (i) The Laurent series for $z \longrightarrow 1/z^2$ about 0 is

$$\cdots + \frac{0}{z^3} + \frac{1}{z^2} + \frac{0}{z} + 0 + 0z + 0z^2 + \cdots = \frac{1}{z^2};$$

    clearly there is no term in $1/z$, and so the residue at 0 is zero.

    (ii) Since $z \longrightarrow 1/(z-1)^2$ is analytic at 0, the residue at 0 is automatically zero (and, of course, the Laurent Series for $z \longrightarrow 1/(z-1)^2$ about 0 is a Taylor series).

    (iii) The Laurent series for $z \longrightarrow e^z/z^n$ about 0 is

$$\frac{1}{z^n}\left(1 + z + \frac{z^2}{2!} + \cdots + \frac{z^{n-1}}{(n-1)!} + \cdots \right),$$

    and so the residue at 0 is $1/(n-1)!$.

(iv) The Laurent series for $z \longrightarrow (\sin z)/z$ about 0 is

$$\frac{1}{z}\left(z - \frac{z^3}{3!} + \cdots\right) = 1 - \frac{z^2}{3!} + \cdots,$$

and so the residue is zero. Actually, this was to be expected since $z \longrightarrow (\sin z)/z$ has a removable singularity at 0, rather than a pole, so the residue is automatically zero.

(v) The Laurent series for $z \longrightarrow z^2 \sin \frac{1}{z}$ about 0 is

$$z^2\left(\frac{1}{z} - \frac{1}{3! \, z^3} + \frac{1}{5! \, z^5} - \cdots\right),$$

and so the residue at 0 is $-\frac{1}{6}$.

2. (i) Let $z = i + h$. Then

$$\frac{1}{z^2 + 1} = \frac{1}{2ih + h^2} = \frac{1}{2ih}\left(1 + \frac{h}{2i}\right)^{-1}$$

$$= \frac{1}{2ih}\left(1 - \frac{h}{2i} + \cdots\right),$$

and so the residue of $z \longrightarrow 1/(z^2 + 1)$ at i is $\frac{1}{2i}$, the coefficient of h^{-1}.

(You will see a better method for this one later in the section.)

(ii) Let $z = \pi + h$. Then

$$\frac{ze^{iz}}{(z - \pi)^2} = \frac{1}{h^2}[(\pi + h)e^{i(\pi + h)}]$$

$$= \frac{\pi + h}{h^2} e^{i\pi} e^{ih}$$

$$= -\frac{\pi + h}{h^2}\left(1 + ih + \frac{(ih)^2}{2!} + \cdots\right), \text{ since } e^{i\pi} = -1.$$

The coefficient of h^{-1} in this expression is $-1 - i\pi$, and so the residue of $z \longrightarrow ze^{iz}/(z - \pi)^2$ at π is $-1 - i\pi$.

(iii) The Laurent series for $z \longrightarrow 1/(z^2 \sin z)$ about 0 is

$$\frac{1}{z^2}\left(z - \frac{z^3}{3!} + \cdots\right)^{-1} = \frac{1}{z^3}\left(1 - \left[\frac{z^2}{3!} - \cdots\right]\right)^{-1}$$

$$= \frac{1}{z^3}\left(1 + \left[\frac{z^2}{3!} - \cdots\right] + \left[\frac{z^2}{3!} - \cdots\right]^2 + \cdots\right)$$

$$= \frac{1}{z^3}\left(1 + \frac{z^2}{3!} + \cdots\right).$$

The coefficient of z^{-1} in this expression is $\frac{1}{6}$, and so the residue of $z \longrightarrow 1/(z^2 \sin z)$ at 0 is $\frac{1}{6}$.

(iv) The residue of $z \longrightarrow (z^2 + 1)^n/z^{n+1}$ at 0 is the coefficient of z^{-1} in $\frac{(z^2 + 1)^n}{z^{n+1}}$, which is the same as the coefficient of z^n in $(z^2 + 1)^n$; this is 0 if n is odd, and $\binom{n}{\frac{1}{2}n}$, the binomial coefficient, if n is even.

3. The result is false, and almost any function will serve as a counter-example. For example, take $f(z) = 1/z$, $\alpha = 0$.

The method described above can become unbelievably complicated if there are too many infinite series to play around with. For example, if we want to calculate the residue at $\frac{1}{2}$ of the function

$$f(z) = \frac{1}{z^3(1-z)(1-2z)(1-3z)^2},$$

then we have to calculate the coefficient of $(z - \frac{1}{2})^{-1}$ in the Laurent series for f about $\frac{1}{2}$, and this is quite a job, as you can imagine. For this reason, it would be useful if we had some short cuts at our disposal. For simple poles, there are some very handy tricks that can be used.

Methods for Simple Poles

Suppose that the function f has a simple pole at α. Then f has a Laurent series of the form

$$f(z) = \frac{a_{-1}}{z - \alpha} + a_0 + a_1(z - \alpha) + \cdots.$$

If we now multiply through by $z - \alpha$, and take the limit as z approaches α, then all but the first term on the right-hand side will disappear, and we get the following result which we have already met in *Unit 8*, Section 8.8):

Theorem 1

If the function f has a simple pole at α, then the residue of f at α, $\text{Res}(f, \alpha)$, is given by

$$\text{Res}(f, \alpha) = \lim_{z \to \alpha} (z - \alpha)f(z).$$

Example 5

Find the residue of $f : z \longrightarrow \dfrac{1}{z^2 + 1}$ at i.

Solution

By Theorem 1,

$$\text{Res}(f, i) = \lim_{z \to i} (z - i) \cdot \frac{1}{z^2 + 1} = \lim_{z \to i} \frac{1}{z + i} = \frac{1}{2i}.$$

You should compare this solution with the solution given for Self-Assessment Question 2(i).

As we have seen in *Unit 8*, Section 8.8, this procedure can be generalized to any rational function with a simple pole, and we get the following result:

The Cover-up Rule

If p and q are polynomials with $p(\alpha) \neq 0$, $q(\alpha) \neq 0$, then the residue of the function

$z \longrightarrow \dfrac{p(z)}{(z - \alpha)q(z)}$ at α is $\dfrac{p(\alpha)}{q(\alpha)}$.

This result can be deduced immediately from Theorem 1, as we saw in Section 8.8 of *Unit 8*.

Example 6

Find the residue of $z \longrightarrow \dfrac{z + 2}{z^3(z + 4)}$ at -4.

Solution

By the cover-up rule, the residue is $\dfrac{-4 + 2}{(-4)^3} = \dfrac{1}{32}$.

The name 'cover-up' rule derives from the fact that what we are effectively doing is covering-up the term $z - \alpha$, and then putting $z = \alpha$ in whatever remains. For instance, in Example 6, we covered up $z + 4$ and put $z = -4$ into what remained.

A generalization of the cover-up rule is the following result, which is possibly the most useful of all rules for evaluating residues at simple poles.

Theorem 2

If $f(z) = \dfrac{g(z)}{h(z)}$, where $g(\alpha) \neq 0, h(\alpha) = 0$, and $h'(\alpha) \neq 0$, then the residue of f at α is equal to $\dfrac{g(\alpha)}{h'(\alpha)}$.

First Proof

The conditions given in the statement of the theorem tell us that f has a simple pole at α. It follows from Theorem 1 that the residue of f at α is given by

$$\text{Res}(f, \alpha) = \lim_{z \to \alpha} (z - \alpha)\frac{g(z)}{h(z)} = \lim_{z \to \alpha} g(z) \cdot \frac{z - \alpha}{h(z) - h(\alpha)} = \frac{g(\alpha)}{h'(\alpha)}.$$

Second Proof

We can use l'Hôpital's Rule (see *Unit 6, Taylor Series*), giving

$$\lim_{z \to \alpha} \frac{(z - \alpha)g(z)}{h(z)} = \lim_{z \to \alpha} \frac{(z - \alpha)g'(z) + g(z)}{h'(z)} = \frac{g(\alpha)}{h'(\alpha)}. \quad \blacksquare$$

Example 7

Find the residue of $f \colon z \longrightarrow \dfrac{z^2 - z + 9}{(z^2 + 9)(z^2 + 1)}$ at $3i$.

First Solution

Let $g(z) = z^2 - z + 9; h(z) = (z^2 + 9)(z^2 + 1)$. Then

$$g(3i) = (3i)^2 - 3i + 9 = -3i;$$

$$h(3i) = 0.$$

Also $h'(z) = 2z \cdot (z^2 + 1) + 2z \cdot (z^2 + 9)$, and so $h'(3i) = -48i$. Hence, by Theorem 2,

$$\text{Res}(f, 3i) = \frac{g(3i)}{h'(3i)} = \frac{1}{16}.$$

We can sometimes simplify our calculations significantly by making a more appropriate choice for g, as the following solution shows:

Second Solution

Let $g(z) = \dfrac{z^2 - z - 9}{z^2 + 1}$; $h(z) = z^2 + 9$. Then

$$g(3i) = \frac{(3i)^2 - 3i + 9}{(3i)^2 + 1} = \frac{3i}{8};$$

$$h(3i) = 0;$$

$$h'(3i) = 6i.$$

Hence, by Theorem 2, $\text{Res}(f, 3i) = \frac{1}{16}$, as before.

Example 4 (again)

Find the residue of $z \longrightarrow \dfrac{\pi}{\sin \pi z}$ at n, where n is an integer.

Second Solution

Let $g(z) = \pi$; $h(z) = \sin \pi z$. Then

$$g(n) = \pi, \quad h(n) = 0 \quad \text{and} \quad h'(n) = \pi \cos \pi n.$$

Hence, by Theorem 2, the residue is

$$\frac{g(n)}{h'(n)} = \frac{\pi}{\pi \cos \pi n} = (-1)^n.$$

Self-Assessment Questions

4. Deduce the cover-up rule from Theorem 2.

5. Find the residues of the following functions f at the points α indicated. (We shall need these residues later in the unit.)

   (i) $f : z \longrightarrow \dfrac{\pi \cos \pi z}{\sin \pi z}, \qquad \alpha = n$ (an integer);

   (ii) $f : z \longrightarrow \dfrac{z^2}{z^4 - 1}, \qquad \alpha = 1$;

   (iii) $f : z \longrightarrow \dfrac{z + 3}{(z - 1)(z - 2)(z + 4)}, \qquad \alpha = 1 \text{ and } 2$;

   (iv) $f : z \longrightarrow \dfrac{2}{z^2 + 4iz - 1}, \qquad \alpha = -i(2 - \sqrt{3})$;

   (v) $f : z \longrightarrow \dfrac{e^z}{\sin z}, \qquad \alpha = 0$;

   (vi) $f : z \longrightarrow \dfrac{z^3}{z^4 + 1}, \qquad \alpha = e^{ik\pi/4} (k = 1, 3, 5 \text{ and } 7)$;

   (vii) $f : z \longrightarrow \dfrac{z^2 - z + 9}{(z^2 + 1)(z^2 + 9)}, \qquad \alpha = i$.

6. Find the residues of the function $z \longrightarrow \dfrac{z^2}{z^4 + 1}$ at the points $e^{i\pi/4}$ and $e^{-i\pi/4}$.

   Do you notice anything about your results?

15

Solutions

4. Let p and q be polynomials with $p(\alpha) \neq 0$, $q(\alpha) \neq 0$, and let

$$f(z) = \frac{p(z)}{(z - \alpha)q(z)}.$$

Then f has a simple pole at α.

In Theorem 2, take $g(z) = p(z)$, $h(z) = (z - \alpha)q(z)$. Then $g(\alpha) \neq 0$, $h(\alpha) = 0$, and $h'(\alpha) = q(\alpha) \neq 0$, and so

$$\text{Res}(f, \alpha) = \frac{g(\alpha)}{h'(\alpha)} = \frac{p(\alpha)}{q(\alpha)}, \quad \text{as required.}$$

5. (i) By Theorem 2, $\text{Res}(f, n) = \dfrac{\pi \cos \pi n}{\pi \cos \pi n} = 1$.

    (ii) By Theorem 2, $\text{Res}(f, 1) = \dfrac{1^2}{4 \cdot 1^3} = \dfrac{1}{4}$.

    (iii) By the cover-up rule:

    $$\text{Res}(f, 1) = \frac{1 + 3}{(1 - 2)(1 + 4)} = -\frac{4}{5};$$

    $$\text{Res}(f, 2) = \frac{2 + 3}{(2 - 1)(2 + 4)} = \frac{5}{6}.$$

    (iv) The given function has poles where $z^2 + 4iz - 1 = 0$, that is when $z = -i(2 \pm \sqrt{3})$; these poles are simple.

    By Theorem 2,

    $$\text{Res}(f, -i(2 - \sqrt{3})) = \frac{2}{-2i(2 - \sqrt{3}) + 4i} = \frac{1}{i\sqrt{3}}.$$

    (v) By Theorem 2, $\text{Res}(f, 0) = \dfrac{e^0}{\cos 0} = 1$.

    (If we had used Theorem 1, we would have had $\lim\limits_{z \to 0} \dfrac{z}{\sin z} = 1$; using Theorem 2, we can avoid having to find limits such as this.)

    (vi) Let $g(z) = z^3$, $h(z) = z^4 + 1$; then $\dfrac{g(z)}{h'(z)} = \dfrac{z^3}{4z^3} = \dfrac{1}{4}$. So the residue of $f = g/h$ at each of the four points $e^{ik\pi/4}$ ($k = 1, 3, 5$ and 7) is equal to $\frac{1}{4}$.

    (vii) Let $g(z) = \dfrac{z^2 - z + 9}{z^2 + 9}$, $h(z) = z^2 + 1$. Then $g(i) = \dfrac{8 - i}{8}$, $h'(i) = 2i$; so that, by Theorem 2, the residue is equal to $\dfrac{8 - i}{16i}$.

6. By Theorem 2, the residue of the function at $e^{i\pi/4}$ is

$$\frac{e^{i\pi/2}}{4e^{3i\pi/4}} = \frac{1}{4}e^{-i\pi/4},$$

and its residue at $e^{-i\pi/4}$ is $\frac{1}{4}e^{i\pi/4}$. Note that $e^{-i\pi/4}$ is the complex conjugate of $e^{i\pi/4}$, and that the residues are also conjugates of each other. (This can be generalized as follows: if f is a function such that $f(z)$ is real whenever z is real, and if the complex conjugates α and $\bar{\alpha}$ are poles of f, then the residues of f at α and $\bar{\alpha}$ are conjugate complex numbers. This fact sometimes simplifies the calculation of residues; you may like to try and prove it if you have time.)

We now show how some of these ideas can be extended to poles of higher order.

Methods for Poles of Higher Order

You have already seen how we can use Laurent series to find the residues of functions at poles of order greater than one. For example, you have already seen how to calculate the residue of $z \longrightarrow \dfrac{\sin z}{z^4}$ at 0, and that of $z \longrightarrow \dfrac{1}{z(z-1)^2}$ at 1 (Examples 2 and 3 above). In fact, in many of the examples you will meet, the quickest method is to substitute $z = \alpha + h$, and find the coefficient of h^{-1}, just as you did in the first part of this section.

Generally speaking, the calculation of residues at poles of order greater than one is not very pleasant. However, there are analogues of Theorems 1 and 2 for residues at poles of order 2, 3, ... ; we shall give the analogue of Theorem 1 here since it is sometimes quite useful, but you should *not* try to memorize it.

Let us start with poles of order two. If the function f has a pole of order two at α, then its Laurent series can be written in the form

$$f(z) = \frac{a_{-2}}{(z-\alpha)^2} + \frac{a_{-1}}{z-\alpha} + a_0 + a_1(z-\alpha) + \cdots, \quad \text{where } a_{-2} \neq 0.$$

We want to isolate the term a_{-1}. Obviously we cannot just multiply through by $z - \alpha$ and take limits, as we did before, since we would have trouble with the first term. We get round this problem by multiplying the Laurent series by $(z-\alpha)^2$, and then differentiating the result, to give

$$\frac{d}{dz}\big(f(z)(z-\alpha)^2\big) = 0 + a_{-1} + 2a_0(z-\alpha) + 3a_1(z-\alpha)^2 + \cdots.$$

Taking the limit near α, we obtain

$$a_{-1} = \lim_{z \to \alpha}\left[\frac{d}{dz}\big(f(z)(z-\alpha)^2\big)\right].$$

The same method can be used for a pole of order $m(>2)$, and the residue is then

$$\frac{1}{(m-1)!}\lim_{z \to \alpha}\left[\frac{d^{m-1}}{dz^{m-1}}\big(f(z)(z-\alpha)^m\big)\right]. \qquad (*)$$

You will have a chance to prove this horrific-looking formula in the next set of Self-Assessment Questions. But first, an example:

Example 8

Find the residue of $f : z \longrightarrow \dfrac{z+2}{z^3(z+4)}$ at 0.

Solution

The function f has a pole of order three at 0, and so the residue there is equal to

$$\frac{1}{2!}\lim_{z \to 0}\left[\frac{d^2}{dz^2}\left(\frac{z+2}{z+4}\right)\right] = \frac{1}{2!}\lim_{z \to 0}\left[\frac{-4}{(z+4)^3}\right] = -\frac{1}{32}.$$

Self-Assessment Questions

7. (a) Prove the above formula $(*)$ for the residue at a pole of order m.

   (b) Use formula $(*)$ to calculate the residues of the following functions at the points α indicated:

   (i) $z \longrightarrow \dfrac{ze^{iz}}{(z-\pi)^2}, \quad \alpha = \pi;$

   (ii) $z \longrightarrow \dfrac{1+e^z}{z^4}, \quad \alpha = 0.$

8. Solve the problem in Example 8 by finding the appropriate Laurent series.

Solutions

7.   (a)   If f has a pole of order m at α, then f can be written in the form

$$f(z) = \frac{a_{-m}}{(z-\alpha)^m} + \frac{a_{-m+1}}{(z-\alpha)^{m-1}} + \cdots + \frac{a_{-1}}{z-\alpha} + a_0$$
$$+ a_1(z-\alpha) + \cdots, \quad a_{-m} \neq 0,$$

and hence

$$f(z)(z-\alpha)^m = a_{-m} + a_{-m+1}(z-\alpha) + \cdots + a_{-1}(z-\alpha)^{m-1}$$
$$+ a_0(z-\alpha)^m + \cdots.$$

Differentiating $(m-1)$ times gives

$$\frac{d^{m-1}}{dz^{m-1}}\big(f(z)(z-\alpha)^m\big) = (m-1)!\,a_{-1} + m!\,a_0(z-\alpha)$$

$$+ \frac{(m+1)!}{2!}a_1(z-\alpha)^2 + \cdots.$$

Dividing by $(m-1)!$ and taking the limit near α gives the required result.

   (b)   (i)   The function has a pole of order two at π, so the residue is

$$\frac{1}{1!}\lim_{z\to\pi}\left(\frac{d}{dz}(ze^{iz})\right) = \lim_{z\to\pi}(e^{iz} + ize^{iz}) = -1 - i\pi.$$

   (ii)   The residue is

$$\frac{1}{3!}\lim_{z\to 0}\left(\frac{d^3}{dz^3}(1+e^z)\right) = \frac{1}{6}\lim_{z\to 0}e^z = \frac{1}{6}.$$

8. $\dfrac{z+2}{z^3(z+4)} = \dfrac{1}{4z^3}(z+2)\left(1+\dfrac{z}{4}\right)^{-1}$

$$= \frac{1}{4z^3}(z+2)\left[1 - \frac{z}{4} + \left(\frac{z}{4}\right)^2 - \cdots\right],$$

and the coefficient of z^{-1} is $\dfrac{1}{4}\left[-\dfrac{1}{4} + 2\left(\dfrac{1}{4}\right)^2\right] = -\dfrac{1}{32}.$

Summary

In this section we discussed the following methods for calculating residues:

(i)   working directly from Laurent series, using, where convenient, the substitution $z = \alpha + h$;

(ii)   the cover-up method, limit method, and 'g/h' method' for simple poles;

(iii)   the formula for poles of order higher than one.

10.2   USING THE RESIDUE THEOREM

In this section, we shall show you how the Residue Theorem can be used to evaluate various types of contour integral. You have already come across some simple examples of this in earlier units—at the end of *Unit 8*, where you considered the special case of a function with just one pole inside a circle, and at the end of *Unit 9*, where the Residue Theorem was used to evaluate a couple of simple integrals.

Our aim here is to continue this line of thought, and to let you try several different types of integral. The basic method will always be the same:

(1)   set up the integral of a suitable function f around a suitable simple-closed contour Γ;

(2)   identify the poles of f which lie *inside* Γ;

(3)   calculate the residues of f at these poles.

It follows from the Residue Theorem that if we add these residues, and multiply the result by $2\pi i$, then we obtain the value of the integral. (If you are not convinced by this, look back at the statement of the Residue Theorem, given in the Introduction. It is essential that you understand the method just described before carrying on.)

Throughout this unit, we shall be concerned with those functions whose only singularities are poles. Without further ado, let us look at an example.

Example 1

Evaluate the integral

$$\int_{\Gamma} \frac{z+3}{(z-1)(z-2)(z+4)}\,dz,$$

where Γ is the ellipse $\dfrac{x^2}{9} + \dfrac{y^2}{4} = 1$.

Solution

To do this problem directly, by parametrizing Γ, is certainly possible, but very tedious. (Try it and see, if you do not believe us!) So let us follow the method outlined above.

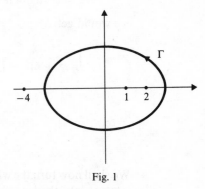

We first identify the poles of the function $z \longrightarrow \dfrac{z+3}{(z-1)(z-2)(z+4)}$. These poles occur at the zeros of the denominator, namely at 1, 2 and -4, and are all simple. Of these, the only poles of the function which lie inside Γ are those at the points 1 and 2 (Fig. 1).

Fig. 1

We have already calculated the residues at these points using the cover-up rule (see Self-Assessment Question 5(iii) on page 15) and they are $-\frac{4}{5}$ (at the point 1) and $\frac{5}{6}$ (at the point 2).

By the Residue Theorem, the value of the integral is

$$2\pi i(\text{sum of these residues}) = 2\pi i\left(-\frac{4}{5} + \frac{5}{6}\right) = \frac{\pi i}{15}.$$

Note that although the function has a pole at -4, we do *not* need to calculate the associated residue since -4 lies outside Γ. Calculating residues at points outside a closed contour is very time-wasting. Be warned!

Example 2

Evaluate the integral

$$\int_\Gamma \frac{z^3}{z^4 + 1}dz,$$

where Γ is the closed semicircular contour shown in Fig. 2.

Solution

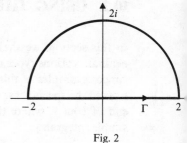

Fig. 2

The function $z \longrightarrow \dfrac{z^3}{z^4 + 1}$ has simple poles at the zeros of $z \longrightarrow z^4 + 1$, that is, at the points $e^{ik\pi/4}$, $k = 1, 3, 5, 7$. Of these poles, only the first two lie inside Γ, and the residue of the function at each of these two points is $\frac{1}{4}$ (see Self-Assessment Question 5(vi) on page 15).

The value of the integral is therefore $2\pi i(\frac{1}{4} + \frac{1}{4}) = \pi i$.

There is one other type of integral that we want to mention before letting you loose on lots of examples. These are the 'real trigonometric integrals', involving sin and cos; for example

$$\int_0^{2\pi} \frac{d\theta}{2 + \sin \theta}, \int_0^{2\pi} \frac{d\theta}{4\cos^2 \theta + \sin^2 \theta} \quad \text{and} \quad \int_0^{2\pi} \frac{\sin \theta}{\cos^2 \theta - \sin \theta}d\theta.$$

(The values of such integrals are, of course, real numbers. This fact will provide a useful check later.) To see how these are related to contour integrals, let us first consider the following contour integral:

$$\int_C \frac{2}{z^2 + 4iz - 1}dz,$$

where C is the unit circle.

If we tried to evaluate this integral by using the substitution

$$z = e^{i\theta}, dz = ie^{i\theta}d\theta \ (0 \leqslant \theta \leqslant 2\pi),$$

we would get

$$\int_C \frac{2}{z^2 + 4iz - 1}dz = \int_0^{2\pi} \frac{2ie^{i\theta}}{e^{2i\theta} + 4ie^{i\theta} - 1}d\theta$$

$$= \int_0^{2\pi} \frac{2i}{e^{i\theta} + 4i - e^{-i\theta}}d\theta$$

$$= \int_0^{2\pi} \frac{d\theta}{2 + \sin \theta}, \quad \text{since } e^{i\theta} - e^{-i\theta} = 2i \sin \theta.$$

We shall now turn the whole argument around, by transforming a trigonometric integral into the 'corresponding' contour integral around C (namely, the one from which it could have been derived using the substitution $z = e^{i\theta}$), and then using the Residue Theorem to evaluate this contour integral. In effect, this means that we replace

$$\cos \theta \text{ by } \frac{1}{2}(z + z^{-1}), \quad \text{since } z + z^{-1} = e^{i\theta} + e^{-i\theta} = 2\cos \theta;$$

$$\sin \theta \text{ by } \frac{1}{2i}(z - z^{-1}), \quad \text{since } z - z^{-1} = e^{i\theta} - e^{-i\theta} = 2i\sin \theta;$$

and

$$d\theta \text{ by } \frac{dz}{iz}, \quad \text{since } \frac{dz}{d\theta} = ie^{i\theta} = iz.$$

An example will make the method clear.

Example 3

Evaluate $\int_0^{2\pi} \dfrac{d\theta}{2 + \sin\theta}$.

Solution

If C is the unit circle, then

$$\int_0^{2\pi} \frac{d\theta}{2 + \sin\theta} = \int_C \frac{1}{2 + \dfrac{1}{2i}(z - z^{-1})} \cdot \frac{dz}{iz}$$

$$= \int_C \frac{2}{z^2 + 4iz - 1} dz, \quad \text{as expected.}$$

The poles of $f : z \longrightarrow \dfrac{2}{z^2 + 4iz - 1}$ occur at the zeros of $z \longrightarrow z^2 + 4iz - 1$, that is, at $-i(2 - \sqrt{3})$ and $-i(2 + \sqrt{3})$. Of these poles, only the first lies inside C, and the residue of f there is $\dfrac{1}{i\sqrt{3}}$ (see Self-Assessment Question 5(iv) on page 15).

The value of the integral is therefore $2\pi i \cdot \dfrac{1}{i\sqrt{3}} = \dfrac{2\pi}{\sqrt{3}}$. (Note, as a check, that the final result is a real number.)

Summary

In this section, we showed how contour integrals can be evaluated by calculating the residues of the poles inside the contour in question. We also showed how the Residue Theorem can be used to evaluate real trigonometric integrals.

Self-Assessment Questions

1. Use the Residue Theorem to deduce that

$$\int_C \frac{dz}{z - \alpha} = 2\pi i,$$

where C is any circle, centre α.

2. Fill in the blanks of the solution to the following problem:

Problem Evaluate the integral

$$\int_\Gamma \frac{z^2 - z + 9}{(z^2 + 1)(z^2 + 9)} dz,$$

where Γ is the rectangle with vertices $-4, 4, 4(1 + i), 4(-1 + i)$.

Solution The poles of the integrand are the points $\boxed{\text{(i)}\qquad\qquad\qquad}$

of which only the two points $\boxed{\text{(ii)}\qquad\qquad}$ lie inside Γ. The residues

at these two points are $\boxed{\text{(iii)}\qquad\qquad}$ and $\boxed{\text{(iv)}\qquad\qquad}$,

respectively, and the value of the integral is therefore $\boxed{\text{(v)}\qquad\qquad}$.

3. Let C be the circle $\{z : |z| = 4\}$. For each of the following functions f, evaluate the integral $\int_C f$. (You have already calculated the relevant residues; see Self-Assessment Question 7 on page 17).

    (i) $f(z) = \dfrac{ze^{iz}}{(z - \pi)^2}$;

    (ii) $f(z) = \dfrac{1 + e^z}{z^4}$.

4. Convert the following trigonometric integral into a contour integral:

    $$\int_0^{2\pi} \frac{d\theta}{4\cos^2\theta + \sin^2\theta}.$$

    (Do not evaluate it: see Problem 3(iii) of Section 10.3.)

Solutions

1. The function $z \longrightarrow \dfrac{1}{z - \alpha}$ has a simple pole at α, and its residue there is 1.
    By the Residue Theorem, the value of the integral is therefore $2\pi i \cdot 1 = 2\pi i$.

2. (i) $+i, -i, +3i, -3i$.

    (ii) $+i, +3i$.

    (iii) $\dfrac{8 - i}{16i}$ (see Self-Assessment Question 5(vii) on page 15).

    (iv) $\frac{1}{16}$ (see Example 7 on page 14).

    (v) $2\pi i\left(\dfrac{8 - i}{16i} + \dfrac{1}{16}\right) = \pi$.

3. (i) f has a pole of order two at π, and its residue there is $-1 - i\pi$. The value of the integral is therefore
    $$2\pi i(-1 - i\pi) = 2\pi^2 - 2\pi i.$$

    (ii) f has a pole of order four at 0, and its residue there is $\frac{1}{6}$. The value of the integral is therefore $\frac{1}{3}\pi i$.

4. Setting $\cos\theta = \dfrac{1}{2}(z + z^{-1})$, $\sin\theta = \dfrac{1}{2i}(z + z^{-1})$, and $d\theta = \dfrac{dz}{iz}$ gives, after some algebra,
    $$\frac{4}{i}\int_C \frac{z}{3z^4 + 10z^2 + 3}dz,$$
    where C is the unit circle.

22

10.3 PROBLEMS

1. Let C be the circle $\{z: |z - i| = 2\}$. Evaluate the integral

    $$\int_C \frac{z + 2}{4z^2 + k^2} dz,$$

    in the following cases:

    (i) $k = 1$;   (ii) $k = 3$;   (iii) $k = 7$.

2. Evaluate the integral

    $$\int_\Gamma \frac{1 + z}{\sin z} dz,$$

    where Γ is the square whose vertices are the points $4 + 4i, -4 + 4i,$ $-4 - 4i, 4 - 4i$.

3. Use the method of the previous section to evaluate the following trigonometric integrals.

    (i) $\displaystyle\int_0^{2\pi} \frac{d\theta}{2 + \cos\theta}.$

    (ii) $\displaystyle\int_0^{2\pi} \cos^n\theta\, d\theta$, where n is a positive integer.

    (For this part you may find it helpful to use the result of Self-Assessment Question 2(iv) on page 11.)

    (iii) $\displaystyle\int_0^{2\pi} \frac{d\theta}{4\cos^2\theta + \sin^2\theta}.$

    (Use the result of Self-Assessment Question 4 on page 22.)

Solutions

1. The function $z \longrightarrow \dfrac{z + 2}{4z^2 + k^2}$ has simple poles at the points $\frac{1}{2}ki$ and $-\frac{1}{2}ki$, and by the cover-up rule its residues there are $\dfrac{\frac{1}{2}ki + 2}{4ki}$ and $\dfrac{-\frac{1}{2}ki + 2}{-4ki}$, respectively.

    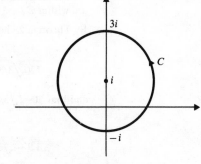

    Fig. 3

    (i) If $k = 1$, then both poles lie inside C (Fig. 3), so that the value of the integral is

    $$2\pi i\left(\frac{\frac{3}{2}i + 2}{12i}\right) = \frac{\pi}{3} + \frac{\pi i}{4}.$$

    (ii) If $k = 3$, the pole at $\frac{3}{2}i$ lies inside C, but the pole at $-\frac{3}{2}i$ lies outside C, so that the value of the integral is

    $$2\pi i\left(\frac{\frac{3}{2}i + 2}{12i}\right) = \frac{\pi}{3} + \frac{\pi i}{4}.$$

    (iii) If $k = 7$, then both poles lie outside C, so that the value of the integral is zero.

2. The poles of the function $z \longrightarrow \dfrac{1 + z}{\sin z}$ occur when $\sin z = 0$, that is, when $z = n\pi$, where n is an integer. It follows that the only poles which lie inside Γ are those at the points $0, \pi$ and $-\pi$, and these are all simple poles. The residues at these points may easily be calculated using Theorem 2, and are $1, -1 - \pi$, and $-1 + \pi$, respectively. The value of the integral is therefore

    $$2\pi i[1 + (-1 - \pi) + (-1 + \pi)] = -2\pi i.$$

    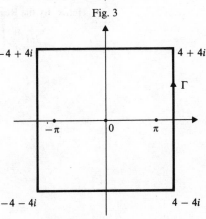

    Fig. 4

3. (i) Setting $\cos\theta = \frac{1}{2}(z + z^{-1})$, $d\theta = \dfrac{dz}{iz}$, we get

$$\int_0^{2\pi} \frac{d\theta}{2 + \cos\theta} = \frac{2}{i}\int_C \frac{dz}{z^2 + 4z + 1},$$

where C is the unit circle.

But $z^2 + 4z + 1 = 0$ when $z = -2 + \sqrt{3}$ and $-2 - \sqrt{3}$, and only the first of these lies inside C. So the only pole of the function $z \longrightarrow \dfrac{1}{z^2 + 4z + 1}$ inside C

is a simple pole at $-2 + \sqrt{3}$, and its residue there is $\dfrac{1}{2\sqrt{3}}$.

The value of the integral is therefore $2\pi i \cdot \dfrac{2}{i} \cdot \dfrac{1}{2\sqrt{3}} = \dfrac{2\pi}{\sqrt{3}}$.

(ii) $$\int_0^{2\pi} \cos^n\theta \, d\theta = \int_C \frac{(z + z^{-1})^n}{2^n} \cdot \frac{dz}{iz}$$

$$= \frac{1}{2^n i}\int_C \frac{(z^2 + 1)^n}{z^{n+1}} \, dz.$$

The function $z \longrightarrow \dfrac{(z^2 + 1)^n}{z^{n+1}}$ has a pole of order $n + 1$ at 0, and by Self-Assessment

Question 2(iv) on page 11, its residue there is 0 if n is odd and $\binom{n}{\frac{1}{2}n}$ if n is even.

Hence, by the Residue Theorem,

$$\int_0^{2\pi} \cos^n\theta \, d\theta = \begin{cases} 0, & \text{if } n \text{ is odd} \\ \dfrac{\pi}{2^{n-1}}\dbinom{n}{\frac{1}{2}n}, & \text{if } n \text{ is even.} \end{cases}$$

(iii) From Self-Assessment Question 4 on page 22.

$$\int_0^{2\pi} \frac{d\theta}{4\cos^2\theta + \sin^2\theta} = \frac{4}{i}\int_C \frac{z}{3z^4 + 10z^2 + 3} \, dz,$$

where C is the unit circle.

Since $3z^4 + 10z^2 + 3 = (z^2 + 3)(3z^2 + 1)$, the function $z \longrightarrow \dfrac{z}{3z^4 + 10z^2 + 3}$

has simple poles at $i\sqrt{3}, -i\sqrt{3}, i\sqrt{3}/3, -i\sqrt{3}/3$. Of these only $i\sqrt{3}/3$ and $-i\sqrt{3}/3$ are within C.

By Theorem 2, the residue at $i\sqrt{3}/3$ is

$$\frac{i\sqrt{3}/3}{12(i\sqrt{3}/3)^3 + 20(i\sqrt{3}/3)} = \frac{1}{16},$$

and that at $-i\sqrt{3}/3$ is

$$\frac{-i\sqrt{3}/3}{12(-i\sqrt{3}/3)^3 + 20(-i\sqrt{3}/3)} = \frac{1}{16}.$$

Hence, by the Residue Theorem, the value of the integral is

$$2\pi i \cdot \frac{4}{i} \cdot \left(\frac{1}{16} + \frac{1}{16}\right) = \pi.$$

10.4 IMPROPER REAL INTEGRALS

Integrals of the form $\int_0^\infty f$ or $\int_{-\infty}^\infty f$ occur in many branches of mathematics, and they are of interest for a variety of reasons. For example, the improper real integral, $\int_{-\infty}^\infty e^{-x^2}\,dx$ is of interest to statisticians because of its relation to the normal distribution. Other examples are provided by the Fourier transform and the Laplace transform of a (suitable) given function F: the Fourier transform of F, denoted by $\tilde{F}$, is defined by

$$\tilde{F}(z) = \frac{1}{\sqrt{2\pi}} \int_{-\infty}^\infty e^{izt}F(t)\,dt,$$

and the Laplace transform of F, denoted by $\mathcal{L}[F]$, is defined by

$$\mathcal{L}[F](z) = \int_0^\infty e^{-zt} F(t)\,dt$$

(note that both the integrands are complex-valued). These two transforms arise in the theory of differential equations, and the latter is discussed in detail in *Unit 14*.

There are good reasons why an integral of the form $\int_0^\infty f$ is easier to deal with than $\int_0^R f$. At a fundamental level, we may consider it an advantage to eliminate a variable, that is R, from our calculations. A more sophisticated reason is that it might be possible to evaluate $\int_0^\infty f$ in terms of the elementary functions (sin, exp, log, etc.), whereas $\int_0^R f$ may be impossible to evaluate in these terms. For instance, $\int_0^\infty e^{-x^2}\,dx = \frac{\sqrt{\pi}}{2}$, whereas there is no simple formula for $\int_0^R e^{-x^2}\,dx$.

The method we shall use to evaluate improper real integrals is to consider the integral of a related (complex) function around a suitable closed contour, which includes part of the real axis. As you will see, there are various ways in which this method can be carried out, but we shall be content with just one or two of the more important methods. Further types of integral will be discussed in the next unit.

Before proceeding, you should work through the following Preliminary Problem, since it will help you to understand the rest of this section more easily.

Preliminary Problem

(a) Use the Residue Theorem to evaluate the integral

$$\int_\Gamma \frac{dz}{1 + z^2},$$

where Γ is the closed semicircular contour of radius $r(>1)$ shown in Fig. 5.

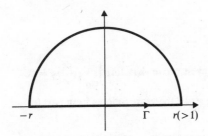

Fig. 5

25

(b) If $\Gamma(r)$ is the semicircular part of Γ (Fig. 6), prove that

$$\left|\frac{1}{1+z^2}\right| \leqslant \frac{1}{r^2-1} \quad \text{for} \quad z \in \Gamma(r).$$

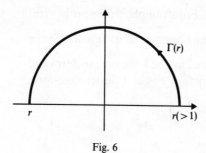

Fig. 6

(c) Use the result of part (b), and the Estimation Theorem to prove that

$$\left|\int_{\Gamma(r)} \frac{dz}{1+z^2}\right| \leqslant \frac{\pi r}{r^2-1},$$

and deduce that $\displaystyle \lim_{r \to \infty} \left|\int_{\Gamma(r)} \frac{dz}{1+z^2}\right| = 0.$

(d) Use the results of parts (a) and (c) to prove that

$$\lim_{r \to \infty} \int_{-r}^{r} \frac{dt}{1+t^2} = \pi,$$

where the integral in question is a *real* integral.

Remark Note that the result of part (d) is what you would expect, since

$$\lim_{r \to \infty} \int_{-r}^{r} \frac{dt}{1+t^2} = \lim_{r \to \infty} \left(\arctan t \Big|_{-r}^{r}\right) = \lim_{r \to \infty} 2 \arctan r = \pi.$$

Solution

(a) The only pole of $z \longrightarrow \dfrac{1}{1+z^2}$ lying inside Γ is a simple pole at i: the residue there is $\dfrac{1}{2i}$. It follows from the Residue Theorem that $\displaystyle \int_{\Gamma} \frac{dz}{1+z^2} = 2\pi i \cdot \frac{1}{2i} = \pi.$

(b) Since $|z| = r$ for z on $\Gamma(r)$, we have

$$|1+z^2| \geqslant r^2 - 1,$$

and the result follows.

(c) The Estimation Theorem states that, if Γ is a contour of length L, and if f is a function whose domain contains Γ, then $\left|\int_{\Gamma} f\right| \leqslant ML$, where M is an upper bound of $|f|$ on Γ.

In our case, $L = \pi r$, and (by part (b)) we can take M to be $\dfrac{1}{r^2-1}$. The result follows immediately. The last part follows from the fact that $\displaystyle \lim_{r \to \infty} \frac{\pi r}{r^2-1} = 0.$

(d) We can split the contour Γ into two parts—the interval $[-r, r]$ of the real axis, which may be trivially parametrized by $\gamma(t) = t$, $-r \leqslant t \leqslant r$, and the semicircular part $\Gamma(r)$. So by part (a),

$$\pi = \int_{\Gamma} \frac{dz}{1+z^2} = \int_{-r}^{r} \frac{dt}{1+t^2} + \int_{\Gamma(r)} \frac{dz}{1+z^2}.$$

In order to prove that $\displaystyle \lim_{r \to \infty} \int_{-r}^{r} \frac{dt}{1+t^2} = \pi$, we must prove that $\displaystyle \lim_{r \to \infty} \int_{\Gamma(r)} \frac{dz}{1+z^2} = 0$, or (equivalently) that $\displaystyle \lim_{r \to \infty} \left|\int_{\Gamma(r)} \frac{dz}{1+z^2}\right| = 0.$ The result therefore follows from part (c).

You should read through this solution several times until you are sure you understand it. In particular, you should make sure you remember exactly where the Residue Theorem was used, and why the Estimation Theorem was introduced. As you can see, what we have done, in effect, is to use techniques of complex analysis to get information about a real integral of the form

$\lim\limits_{r \to \infty} \int_{-r}^{r} f(t)\, dt$, by means of the following steps:

General Procedure	Particular Problem
(1) We associated with the given real integral, a related contour integral, of the form $\int_{\Gamma} f(z)\, dz$.	(1) Consider $\int_{\Gamma} \dfrac{dz}{1 + z^2}$.
(2) We used the Residue Theorem to evaluate the contour integral.	(2) $\int_{\Gamma} \dfrac{dz}{1 + z^2} = \pi$.
(3) We split the contour integral into two parts: a real integral we are interested in, and a complex integral we want to get rid of.	(3) $\int_{\Gamma} \dfrac{dz}{1 + z^2} = \int_{-r}^{r} \dfrac{dt}{1 + t^2} + \int_{\Gamma(r)} \dfrac{dz}{1 + z^2}$.
(4) We used the Estimation Theorem to show that this complex integral becomes arbitrarily small in modulus if we take r to be large enough.	(4) $\lim\limits_{r \to \infty} \left| \int_{\Gamma(r)} \dfrac{dz}{1 + z^2} \right| = 0$.

Our main purpose is to show how these steps can be adapted to evaluate a whole class of real integrals of the form $\int_{-\infty}^{\infty} f$. The basic idea will be to define this to be $\lim\limits_{r \to \infty} \int_{-r}^{r} f$, and to convert this into a contour integral of the form $\int_{\Gamma} f$ by 'closing up the contour', with a semicircular (or similar) arc; the resulting contour integral is then evaluated using the Residue Theorem. Before you see how this works in practice, we must give an important definition.

Definition

> If f is a complex-valued function of a real variable, then we define the **improper integral** $\int_{-\infty}^{\infty} f(t)\, dt$ to be the number defined by the formula
>
> $$\int_{-\infty}^{\infty} f(t)\, dt = \lim_{r \to \infty} \int_{-r}^{r} f(t)\, dt,$$
>
> provided that this limit exists. If the limit exists, we also say that the integral $\int_{-\infty}^{\infty} f(t)\, dt$ **converges**.

Important Notes

(i) This definition is a little different from the definition of improper integral given in **Spivak**, page 254, and in *Unit M231 11, Techniques of Integration*.

(ii) Some textbooks write PV $\int_{-\infty}^{\infty} f(t)\, dt$. (PV stands for 'principal value.')

Example 1

Evaluate   (i) $\int_{-\infty}^{\infty} t\, dt$;   (ii) $\int_{-\infty}^{\infty} \dfrac{dt}{1 + t^2}$.

Solution

(i) $\displaystyle\int_{-\infty}^{\infty} t\,dt = \lim_{r\to\infty}\int_{-r}^{r} t\,dt = \lim_{r\to\infty}\left(\tfrac{1}{2}t^2\Big|_{-r}^{r}\right) = 0.$

(ii) $\displaystyle\int_{-\infty}^{\infty} \frac{dt}{1+t^2} = \lim_{r\to\infty}\int_{-r}^{r} \frac{dt}{1+t^2} = \lim_{r\to\infty}\left(\arctan t\Big|_{-r}^{r}\right) = \pi.$

(This is the example we had before.)

It is a simple matter to prove that if f is an *odd* function, then the integral $\displaystyle\int_{-\infty}^{\infty} f(t)\,dt$ converges, and equals zero. You will be asked to prove this in the next set of self-assessment questions.

We turn now to the evaluation of improper integrals, starting with rational functions.

Theorem 3

Let p and q be polynomials with the following properties:

(i) q has no zeros on the real axis;

(ii)   the degree of q exceeds that of p by at least two.

Then the integral $\displaystyle\int_{-\infty}^{\infty} \frac{p(t)}{q(t)}\,dt$ converges, and equals $2\pi i S$, where S is the sum of the residues of the function $z \longrightarrow \dfrac{p(z)}{q(z)}$ at those poles which lie in the upper half-plane.

Before we prove this result, let us look back at our Preliminary Problem, where we evaluated $\displaystyle\int_{-\infty}^{\infty} \frac{dt}{1+t^2}$. In that case, $p(t) = 1$, $q(t) = 1 + t^2$, so that conditions (i) and (ii) were satisfied, and S turned out to be the residue at the point i—in other words, the sum of the residues of $z \longrightarrow \dfrac{1}{1+z^2}$ at the poles in the upper half-plane. The solution was obtained by considering $\displaystyle\int_{\Gamma} \frac{dz}{1+z^2}$, where Γ was a semicircular contour in the upper half-plane, and it seems at least plausible that the same method will work more generally. So let us try the proof.

Proof of Theorem 3

In the light of the above discussion, let us consider the contour integral

$$\int_{\Gamma} \frac{p(z)}{q(z)}\,dz,$$

where Γ is the semicircular contour shown in Fig. 7.

Since we shall eventually let r become arbitrarily large, we shall assume right from the start that r is large enough for the semicircle to contain all the poles of p/q which lie in the upper half-plane. (This is certainly possible, since there is only a finite number of such poles; in the Preliminary Problem, we took $r > 1$ for the same reason.)

So all of the poles of p/q in the upper half-plane lie inside Γ, and none lie on its boundary, by condition (i); we can therefore apply the Residue Theorem, which tells us that the value of the contour integral is simply $2\pi i S$.

It follows, on splitting Γ into its components, that

$$2\pi i S = \int_{\Gamma} \frac{p(z)}{q(z)}\,dz = \int_{-r}^{r} \frac{p(t)}{q(t)}\,dt + \int_{\Gamma(r)} \frac{p(z)}{q(z)}\,dz,$$

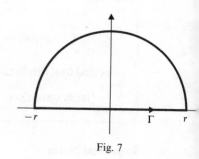

Fig. 7

where $\Gamma(r)$ is the semicircular part of Γ. We shall have proved the theorem if we can prove that this last integral along the semicircle $\Gamma(r)$ has limit zero as r becomes large. To do this, we use the Estimation Theorem, with $L = \pi r$, and a suitable upper bound M for $\left|\dfrac{p}{q}\right|$ on $\Gamma(r)$.

Since the degree of q exceeds that of p by at least two, it seems reasonable to expect that, if r is large enough (that is, $r \geqslant r_0$, say), then

$$\left|\frac{p(z)}{q(z)}\right| \leqslant \frac{K}{r^2} \text{ for } z \text{ on } \Gamma(r), \tag{*}$$

where K is constant. Assuming this for the time being, we then have, by the Estimation Theorem,

$$\left|\int_{\Gamma(r)} \frac{p(z)}{q(z)} dz\right| \leqslant \pi r \cdot \frac{K}{r^2}, \quad \text{if } r \geqslant r_0.$$

It follows that $\displaystyle\lim_{r \to \infty} \int_{\Gamma(r)} \frac{p(z)}{q(z)} dz = 0$, and hence that

$$2\pi i S = \lim_{r \to \infty} \int_{-r}^{r} \frac{p(t)}{q(t)} dt = \int_{-\infty}^{\infty} \frac{p(t)}{q(t)} dt, \quad \text{as required.} \quad \blacksquare$$

The only part of the proof which we skipped over was the inequality marked (*). We shall, in fact, prove a slightly more general result, which will be useful later on.

Theorem 4

Let p and q be polynomials, and suppose that the degree of q exceeds that of p by d. Then there are constants K and r_0, such that, if $|z| \geqslant r_0$, then $\left|\dfrac{p(z)}{q(z)}\right| \leqslant \dfrac{K}{|z|^d}$.

Proof

If p has degree m and q has degree n, where $n - m = d$, then we can write $\dfrac{p(z)}{q(z)}$ in the form

$$\frac{p(z)}{q(z)} = z^{-d} \cdot \frac{a_m + a_{m-1}z^{-1} + \cdots + a_1 z^{1-m} + a_0 z^{-m}}{b_n + b_{n-1}z^{-1} + \cdots + b_1 z^{1-n} + b_0 z^{-n}} \quad (a_m \neq 0, b_n \neq 0)$$

$$= z^{-d} g(z), \text{ say.}$$

Since $\lim_{z \to \infty} g(z)$ is clearly equal to a_m/b_n, there exists a number r_0 such that $|g(z) - a_m/b_n| < 1$, if $|z| \geqslant r_0$.

It follows that, if $|z| \geqslant r_0$, then

$$\left|\frac{p(z)}{q(z)}\right| \leqslant |z^{-d}| \cdot \left(\left|\frac{a_m}{b_n}\right| + 1\right) = \frac{K}{|z|^d}, \quad \text{where } K = \left|\frac{a_m}{b_n} + 1\right|. \quad \blacksquare$$

The proof of Theorem 3 is now complete.

The method of proof of Theorem 3 is important, and we shall show later in the section how it can be adapted to deal with functions other than rational functions; we shall also show how the restrictions imposed by conditions (i) and (ii) can be removed.

Self-Assessment Questions

1. Look back at the proof of Theorem 3, and answer the following questions.

   (i) Where were conditions (i) and (ii) of Theorem 3 used?

   (ii) Why did we use the Estimation Theorem?

   (iii) Why did we need to prove Theorem 4 (or a special case of it)?

   (iv) Why did we start by taking r so large that all of the poles of p/q in the upper half-plane were inside the contour Γ?

   (v) Do you think it would have made any difference if we had closed up the contour with a semicircle in the *lower* half-plane?

   (vi) More generally, could we have closed up the contour with a different type of contour, such as a semi-elliptical or a rectangular contour?

   (vii) Does the integral of p/q along the semicircle $\Gamma(r)$ *have* to approach zero as r becomes large, for the method to work? Could it approach a finite limit l instead?

2. Fill in the blanks in the solution to the following problem:

   Problem   Evaluate $\displaystyle\int_{-\infty}^{\infty} \frac{t^2}{(t^2 + 4)^2}\, dt$.

   Solution   Consider the integral $\boxed{\text{(i)}\hspace{3cm}}$, where Γ is the usual closed semicircular contour. The only pole inside Γ is the pole of order $\boxed{\text{(ii)}\hspace{2.5cm}}$ at the point $\boxed{\text{(iii)}\hspace{2cm}}$, and the residue of the function there is $\dfrac{1}{8i}$.

   Hence

   $$\int_{-r}^{r} \frac{t^2}{(t^2 + 4)^2}\, dt + \int_{\Gamma(r)} \frac{z^2}{(z^2 + 4)^2}\, dz = \boxed{\text{(iv)}\hspace{2.5cm}}.$$

   But by Theorem 4 and the Estimation Theorem, there exist constants r_0 and K such that, if $r \geqslant r_0$, then

   $$\left| \int_{\Gamma(r)} \frac{z^2}{(z^2 + 4)^2}\, dz \right| \leqslant \boxed{\text{(v)}\hspace{2.5cm}},$$

   and this $\boxed{\text{(vi)}\hspace{2.5cm}}$ as r becomes large. The value of the required integral is therefore $\boxed{\text{(vii)}\hspace{2.5cm}}$.

3. (i) If f is an odd function of a real variable, show that the integral $\displaystyle\int_{-\infty}^{\infty} f$ converges, and equals zero.

   (ii) If f is an even function of a real variable, show that

   $$\int_{-\infty}^{\infty} f = 2\int_{0}^{\infty} f,$$

   where $\displaystyle\int_{0}^{\infty} f$ is defined to be $\displaystyle\lim_{r \to \infty} \int_{0}^{r} f$, as in M231, Analysis, and the limit is assumed to exist.

4. (i) Use Theorem 3 to show that

$$\int_{-\infty}^{\infty} \frac{dt}{t^4 + 1} = \frac{\pi\sqrt{2}}{2}.$$

(The residues at $e^{\pi i/4}$ and $e^{3\pi i/4}$ are $\frac{1}{4}e^{-3\pi i/4}$ and $\frac{1}{4}e^{-\pi i/4}$, respectively.)

(ii) Evaluate $\displaystyle\int_{-\infty}^{\infty} \frac{t}{t^4 + 1}\, dt$. (No calculation required!)

Solutions

1. (i) Condition (i) was used when we applied the Residue Theorem to $\int_{\Gamma} \frac{p(z)}{q(z)}\, dz$, since the Residue Theorem requires that the contour does not pass through any poles of the function we are integrating. Condition (ii) was used to find an upper bound of $|p/q|$ on $\Gamma(r)$; if the degree of q had been less than two more than the degree of p, then we could not have proved that the bound given by the Estimation Theorem has limit zero as r becomes large.

(ii) The Estimation Theorem was used to show that the integral of p/q along $\Gamma(r)$ has limit zero as r becomes large. In particular, we used it to prove that

$$\left| \int_{\Gamma(r)} \frac{p(z)}{q(z)}\, dz \right| \leqslant \frac{K\pi}{r}, \text{ if } r \text{ is large.}$$

(iii) We needed an upper bound of $|p/q|$ on $\Gamma(r)$ in order to use the Estimation Theorem; Theorem 4 gives us a particularly simple upper bound.

(iv) By taking r suitably large to begin with, we can guarantee that $\int_{\Gamma} \frac{p(z)}{q(z)}\, dz = 2\pi i S$, where S is the sum of the residues at *all* the poles in the upper half-plane. Consequently, when we increase r indefinitely, no new poles are introduced, so that $2\pi i S$ remains unchanged.

(v) There is no reason at all why we should not use the lower half-plane. Everything will go through much as before. (See also the next answer.)

(vi) Theoretically there is no special virtue in using a semicircular contour for this problem, and, in fact we shall see some problems in which a rectangular contour, or some other type of contour, is more appropriate. All we need is that the integral along our chosen contour behaves suitably (for example, has limit zero) as r becomes large, and we usually choose a semicircle because the corresponding estimation is somewhat simpler. In general, we choose the type of contour which makes the particular problem we are dealing with easiest to handle, and this will, of course, vary from problem to problem.

(vii) In this particular problem, the integral of p/q along $\Gamma(r)$ must have limit zero because of condition (ii). On the other hand, there are problems in which the integral along $\Gamma(r)$ approaches some other finite limit l, in which case the calculation proceeds as before, but with $2\pi i S - l$ instead of $2\pi i S$.

2. (i) $\displaystyle\int_\Gamma \frac{z^2}{(z^2+4)^2}\,dz.$

   (ii) two.

   (iii) $2i$.

   (iv) $\displaystyle 2\pi i\cdot\frac{1}{8i}=\frac{\pi}{4}.$

   (v) $\displaystyle \pi r\cdot\frac{K}{r^2}=\frac{K\pi}{r}.$

   (vi) has limit zero.

   (vii) $\dfrac{\pi}{4}$.

3. (i)
$$\int_{-\infty}^{\infty} f(t)\,dt = \lim_{r\to\infty}\int_{-r}^{r} f(t)\,dt$$
$$= \lim_{r\to\infty}\left(\int_0^r f(t)\,dt + \int_{-r}^0 f(t)\,dt\right)$$
$$= \lim_{r\to\infty}\left(\int_0^r f(t)\,dt + \int_0^r f(-u)\,du\right),\quad \text{putting } u=-t,$$
$$= \lim_{r\to\infty}\left(\int_0^r f(t)\,dt - \int_0^r f(u)\,du\right),\quad \text{since } f \text{ is odd,}$$
$$= 0.$$

   (ii)
$$\int_{-\infty}^{\infty} f(t)\,dt = \lim_{r\to\infty}\left(\int_0^r f(t)\,dt + \int_0^r f(-u)\,du\right),\quad \text{as before,}$$
$$= \lim_{r\to\infty}\left(\int_0^r f(t)\,dt + \int_0^r f(u)\,du\right),\quad \text{since } f \text{ is even,}$$
$$= 2\lim_{r\to\infty}\int_0^r f(t)\,dt = 2\int_0^{\infty} f(t)\,dt.$$

4. (i) Conditions (i) and (ii) are satisfied, so we can apply Theorem 3. The poles of $z \longrightarrow \dfrac{1}{z^4+1}$ occur when $z^4=-1$, that is at the points $e^{k\pi i/4}$, $k=1,3,5,7$, of which only $e^{\pi i/4}$ and $e^{3\pi i/4}$ lie in the upper half-plane. The residues at these points are $\frac14 e^{-3\pi i/4}$ and $\frac14 e^{-\pi i/4}$, respectively, and so, by Theorem 3,

$$\int_{-\infty}^{\infty}\frac{dt}{t^4+1} = 2\pi i\cdot\tfrac14(e^{-3\pi i/4}+e^{-\pi i/4})$$
$$= 2\pi i\cdot\tfrac14(e^{\pi i/4}\cdot e^{-\pi i}+e^{-\pi i/4})$$
$$= 2\pi i\cdot\frac14\cdot\left(-2i\sin\frac{\pi}{4}\right)$$
$$= \frac{\pi\sqrt2}{2}.$$

   (ii) By part (i) of Self-Assessment Question 3, the integral is zero.

For the rest of this section we shall show how the methods and results just described can be modified to deal with some special classes of functions which are not rational functions. The basic method will remain the same, but some of the techniques we use will need to be a little more subtle.

Our first variation will be to look at integrals of the form

$$\int_{-\infty}^{\infty} \frac{p(t)}{q(t)} \cos \beta t \, dt \quad \text{and} \quad \int_{-\infty}^{\infty} \frac{p(t)}{q(t)} \sin \beta t \, dt;$$

where p and q are polynomials, and $\beta > 0$. We start with an example.

Example 2

Evaluate $\int_{-\infty}^{\infty} \frac{\cos t}{t^2 + 4} \, dt$.

Solution

To evaluate this integral using the Residue Theorem, it would seem sensible to consider the integral

$$\int_{\Gamma} \frac{\cos z}{z^2 + 4} \, dz,$$

where Γ is the usual closed semicircular contour in the upper half-plane.

Unfortunately, this does not work at all well, since there is no convenient way of bounding the function $z \longrightarrow \dfrac{\cos z}{z^2 + 4}$ on the semicircle $\Gamma(r)$. One standard way of dealing with this problem is to notice that $\cos t = \text{Re}(e^{it})$, and to consider instead the integral

$$\int_{\Gamma} \frac{e^{iz}}{z^2 + 4} \, dz,$$

where Γ is the usual closed semicircular contour in the upper half-plane.

If $r > 2$, then the function $z \longrightarrow \dfrac{e^{iz}}{z^2 + 4}$ has exactly one singularity inside Γ, namely, a simple pole at the point $2i$, with residue $\dfrac{e^{-2}}{4i}$ (by Theorem 2). It then follows from the Residue Theorem that

$$2\pi i \cdot \frac{e^{-2}}{4i} = \int_{\Gamma} \frac{e^{iz}}{z^2 + 4} \, dz = \int_{-r}^{r} \frac{e^{it}}{t^2 + 4} \, dt + \int_{\Gamma(r)} \frac{e^{iz}}{z^2 + 4} \, dz.$$

Our aim is now to estimate this last integral, and to show that it has limit zero as r becomes large. But if $z = x + iy$, then

$$|e^{iz}| = |e^{ix}e^{-y}| = |e^{ix}\|e^{-y}| = e^{-y} \leqslant 1,$$

since $y \geqslant 0$. (This is why we must use the *upper* half-plane when dealing with integrals involving $\cos t$ or $\sin t$.) Hence, by Theorem 4, there are constants K and r_0 such that, if $r \geqslant r_0$, then

$$\left| \int_{\Gamma(r)} \frac{e^{iz}}{z^2 + 4} \, dz \right| \leqslant \pi r \cdot \frac{K}{r^2}, \quad \text{by the Estimation Theorem.}$$

It follows that the last integral has limit zero as r becomes large, and we get

$$\frac{\pi}{2e^2} = \int_{-\infty}^{\infty} \frac{e^{it}}{t^2 + 4} \, dt.$$

On equating real and imaginary parts, we get

$$\int_{-\infty}^{\infty} \frac{\cos t}{t^2 + 4} \, dt = \frac{\pi}{2e^2}, \quad \text{and} \quad \int_{-\infty}^{\infty} \frac{\sin t}{t^2 + 4} \, dt = 0.$$

(Note that the result of the second integral is what you would expect, since $t \longrightarrow \dfrac{\sin t}{t^2 + 4}$ is an odd function.)

In this last example, the presence of the e^{iz} term made no difference to the estimation of the integral along $\Gamma(r)$, since $|e^{iz}| \leqslant 1$ for $z \in \Gamma(r)$. It follows that integrals of this type, involving $\cos \beta t$ or $\sin \beta t$ are really no more difficult to deal with than the integrals of rational functions. Since the method is exactly the same as before, we should expect that a result of the following type ought to hold, as indeed it does.

Theorem 5

Let p and q be polynomials with the following properties:

(i) q has no zeros on the real axis;

(ii) the degree of q exceeds that of p by at least two.

Then, if $\beta \geqslant 0$,

$$\int_{-\infty}^{\infty} \frac{p(t)}{q(t)} \cos \beta t \, dt = \text{Re}(2\pi i S)$$

and

$$\int_{-\infty}^{\infty} \frac{p(t)}{q(t)} \sin \beta t \, dt = \text{Im}(2\pi i S),$$

where S is the sum of the residues of the function $z \longrightarrow \dfrac{p(z)}{q(z)} e^{i\beta z}$ at those poles which lie in the upper half-plane.

We shall not prove this result here for two reasons:

(1) the proof is almost exactly the same as that of Theorem 3, except for the fact that we now start by considering the integral

$$\int_{\Gamma} \frac{p(z)}{q(z)} e^{i\beta z} \, dz,$$

where Γ is the usual closed semicircular contour in the upper half-plane;

(2) we shall in any case be proving a stronger result (Theorem 7) which will enable us to deal with integrals of the above type where the degree of q is only *one* greater than the degree of p. To do this, we shall make use of a method which is rather more subtle than simply putting $|e^{iz}| \leqslant 1$, as we did before. The method we shall use is based on a result known as Jordan's Lemma.

Theorem 6   (Jordan's Lemma)

Let $\Gamma(r)$ be the semicircular path of radius r parametrized by $\gamma(\theta) = re^{i\theta}$, $0 \leqslant \theta \leqslant \pi$, and let f be a function continuous on $\Gamma(r)$. If $|f| \leqslant M$ on $\Gamma(r)$, and if $\beta > 0$, then

$$\left| \int_{\Gamma(r)} f(z) e^{i\beta z} \, dz \right| < \frac{M\pi}{\beta}.$$

Remark   Note that our former estimate was $M\pi r$, so this present result is a substantial improvement. However, the theorem does not apply if $\beta = 0$, so we cannot use it to improve on the result of Theorem 3.

Proof of Theorem 6

If $z = re^{i\theta} = r \cos \theta + ir \sin \theta$, then

$$e^{i\beta z} = e^{-\beta r \sin \theta} \cdot e^{i\beta r \cos \theta},$$

so that

$$\int_{\Gamma(r)} f(z) e^{i\beta z} \, dz = \int_{0}^{\pi} f(re^{i\theta}) e^{-\beta r \sin \theta} \cdot e^{i\beta r \cos \theta} \cdot rie^{i\theta} \, d\theta.$$

The idea is now to estimate this integral, replacing $|f(re^{i\theta})|$ by M, and $|i|$, $|e^{i\theta}|$ and $|e^{i\beta r \cos\theta}|$ by 1.

We then get

$$\left| \int_{\Gamma(r)} f(z)e^{i\beta z}\, dz \right| \leqslant \int_0^\pi |f(re^{i\theta})| \cdot |e^{-\beta r \sin\theta}| \cdot |e^{i\beta r \cos\theta}| \cdot |rie^{i\theta}|\, d\theta$$

$$\leqslant Mr \int_0^\pi e^{-\beta r \sin\theta}\, d\theta.$$

(Note that we do not need the modulus signs in this last integral, since the integrand is always positive.)

Our problem now is to estimate this last integral; the estimate will have to be a good one, since we need to get rid of the factor r which appears just before the integral. To do this, we use an elementary result in trigonometry, namely that

$$\text{if } 0 \leqslant \theta \leqslant \frac{\pi}{2}, \text{ then } \sin\theta \geqslant \frac{2\theta}{\pi}.$$

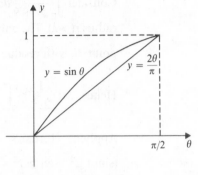

Fig. 8

(This result is clear geometrically, as you can see from Fig. 8. We shall ask you to prove it analytically in Problem 1 of Section 10.5.)

We first show that

$$\int_0^\pi e^{-\beta r \sin\theta}\, d\theta = 2 \int_0^{\pi/2} e^{-\beta r \sin\theta}\, d\theta.$$

We have

$$\int_0^\pi e^{-\beta r \sin\theta}\, d\theta = \int_0^{\pi/2} e^{-\beta r \sin\theta}\, d\theta + \int_{\pi/2}^\pi e^{-\beta r \sin\theta}\, d\theta$$

$$= \int_0^{\pi/2} e^{-\beta r \sin\theta}\, d\theta + \int_{\pi/2}^\pi e^{-\beta r \sin(\pi - \theta)}\, d\theta$$

$$= \int_0^{\pi/2} e^{-\beta r \sin\theta}\, d\theta + \int_0^{\pi/2} e^{-\beta r \sin\phi}\, d\phi,$$

$$\text{where } \phi = \pi - \theta,$$

$$= 2 \int_0^{\pi/2} e^{-\beta r \sin\theta}\, d\theta.$$

Hence,

$$\int_0^\pi e^{-\beta r \sin\theta}\, d\theta \leqslant 2 \int_0^{\pi/2} e^{-2\beta r\theta/\pi}\, d\theta, \quad \text{using the above trigonometric result,}$$

$$= -\frac{\pi}{\beta r} e^{-2\beta r\theta/\pi} \Big|_0^{\pi/2}$$

$$= \frac{\pi}{\beta r}(1 - e^{-\beta r}) < \frac{\pi}{\beta r}.$$

Hence,

$$\left| \int_\Gamma f(z)e^{i\beta z}\, dz \right| < Mr \cdot \frac{\pi}{\beta r} = \frac{M\pi}{\beta}, \quad \text{as required.} \quad \blacksquare$$

You are not expected to remember all of the details of this proof. We have included them because we wanted to show how we can substantially improve our estimation of integrals by taking a bit more care with the trigonometric terms. You should, however, remember the final result, since it will be used several times in what follows.

Before giving the 'improved version' of Theorem 5, let us look at an example in which Jordan's Lemma is used. We shall take an example in which the degree of the polynomial in the denominator is only one greater than that of the numerator.

Example 3

Evaluate $\int_{-\infty}^{\infty} \dfrac{t \sin t}{t^2 + 9} \, dt$.

Solution

Consider $\int_{\Gamma} \dfrac{z e^{iz}}{z^2 + 9} \, dz$, where Γ is the usual closed semicircular contour of radius $r (> 3)$. The only singularity in the upper half-plane is a simple pole at the point $3i$, with residue $\dfrac{3i e^{-3}}{6i} = \dfrac{1}{2e^3}$.

Hence, $2\pi i \cdot \dfrac{1}{2e^3} = \int_{-r}^{r} \dfrac{t e^{it}}{t^2 + 9} \, dt + \int_{\Gamma(r)} \dfrac{z e^{iz}}{z^2 + 9} \, dz$.

Applying Jordan's Lemma with $\beta = 1, f(z) = \dfrac{z}{z^2 + 9}$, and $M = \dfrac{r}{r^2 - 9}$ (since $|z^2 + 9| \geqslant r^2 - 9$), we get

$$\left| \int_{\Gamma(r)} \dfrac{z e^{iz}}{z^2 + 9} \, dz \right| < \dfrac{\pi r}{r^2 - 9},$$

which has limit zero as r becomes large.

It follows that

$$2\pi i \cdot \dfrac{1}{2e^3} = \int_{-\infty}^{\infty} \dfrac{t e^{it}}{t^2 + 9} \, dt,$$

and hence, on taking imaginary parts, that

$$\int_{-\infty}^{\infty} \dfrac{t \sin t}{t^2 + 9} \, dt = \dfrac{\pi}{e^3}.$$

Let us now prove the general theorem. Because you already know how the proof goes, we have left some blanks for you to fill in.

Theorem 7

Let p and q be polynomials with the following properties:

(i) q has no zeros on the real axis;

(ii) the degree of q exceeds that of p by at least one.

Then, if $\beta > 0$,

$$\int_{-\infty}^{\infty} \dfrac{p(t)}{q(t)} \cos \beta t \, dt = \operatorname{Re}(2\pi i S)$$

and

$$\int_{-\infty}^{\infty} \dfrac{p(t)}{q(t)} \sin \beta t \, dt = \operatorname{Im}(2\pi i S),$$

where S is the sum of the residues of the function $z \longrightarrow \dfrac{p(z)}{q(z)} e^{i\beta z}$ at those poles which lie in the upper half-plane.

(Note that the result does not hold if $\beta = 0$.)

Proof

Consider $\displaystyle\int_\Gamma \boxed{\text{(i)}} \; dz,$

in which Γ is the usual closed semicircular contour of radius r, where r is large enough so that

$$\boxed{\text{(ii)}}.$$

By the Residue Theorem, the value of this integral is $2\pi i S$. It follows that

$$2\pi i S = \int_{-r}^{r} \boxed{\text{(iii)}} \; dt + \int_{\Gamma(r)} \boxed{\text{(iv)}} \; dz,$$

where $\Gamma(r)$ is the semicircular path

$$\left\{ z: \boxed{\text{(v)}} \right\}.$$

Applying Jordan's Lemma and Theorem 4 to the last integral gives, for suitable constants K and r_0,

$$\left| \int_{\Gamma(r)} \boxed{\text{(vi)}} \; dz \right| < \boxed{\text{(vii)}}, \quad \text{if } r \geqslant r_0,$$

and this $\boxed{\text{(viii)}}$ as r becomes large.

It follows that

$$2\pi i S = \int_{-\infty}^{\infty} \boxed{\text{(ix)}} \; dt,$$

and the result follows by $\boxed{\text{(x)}}$. ∎

Solutions

(i) $\dfrac{p(z)}{q(z)} e^{i\beta z}.$

(ii) all the poles of $z \longrightarrow \dfrac{p(z)}{q(z)} e^{i\beta z}$ in the upper half-plane lie inside Γ.

(iii) $\dfrac{p(t)}{q(t)} e^{i\beta t}.$

(iv) $\dfrac{p(z)}{q(z)} e^{i\beta z}.$

(v) $z = r e^{i\theta}, \quad 0 \leqslant \theta \leqslant \pi.$

(vi) $\dfrac{p(z)}{q(z)} e^{i\beta z}.$

(vii) $\dfrac{K}{r} \cdot \dfrac{\pi}{\beta}.$

(viii) has limit zero.

(ix) $\dfrac{p(t)}{q(t)} e^{i\beta t}.$

(x) equating real and imaginary parts.

We conclude this section by discussing the integration of functions which have poles on the real axis. Since the Residue Theorem will not work directly in this case, we shall have to 'jiggle the contour a little', so that it avoids the troublesome poles. In what follows, we shall assume that *all poles lying on the real axis are simple poles*.

Before giving an example, we shall need a definition.

Definition

> If f is a complex-valued function of a real variable, and if f is continuous on an interval $[a, b]$ except at the point c, then the **improper integral** $\int_a^b f(t)\, dt$ is the number defined by the formula
>
> $$\int_a^b f(t)\, dt = \lim_{\varepsilon \to 0^+} \left(\int_a^{c-\varepsilon} f(t)\, dt + \int_{c+\varepsilon}^b f(t)\, dt \right)$$
>
> provided that this limit exists.

This definition can clearly be extended to functions f which have discontinuities at a finite number of points on the real axis, and the improper integral $\int_{-\infty}^{\infty} f(t)\, dt$ for such functions is also defined in the obvious way.

In order to integrate a function which has poles on the real axis, we use the usual closed semicircular contour, except that we make small semicircular indentations of radius ε to avoid those points which cause the trouble (Fig. 9). If we now let r become arbitrarily large and ε arbitrarily small, then the integral along the semicircle $\Gamma(r)$ has limit zero and the integral along an indentation of radius ε (taken *clockwise*, as shown in Fig. 9) has limit

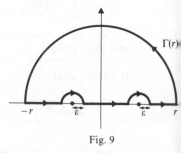

Fig. 9

$-\pi i$ times the residue at the pole in question.

We shall not prove this result here, but will leave it to you to prove in the next problems section. Assuming the result for the time being, it is a straightforward matter to derive the following analogues of Theorems 5 and 7, which will enable you to integrate functions with poles on the real axis.

Theorem 8(a)

Let p and q be polynomials with the following properties:

(i) q has simple zeros on the real axis;

(ii)   the degree of q exceeds that of p by at least two.

Then, if $\beta \geqslant 0$,

$$\int_{-\infty}^{\infty} \frac{p(t)}{q(t)} \cos \beta t\, dt = \operatorname{Re}(2\pi i S + \pi i T)$$

and

$$\int_{-\infty}^{\infty} \frac{p(t)}{q(t)} \sin \beta t\, dt = \operatorname{Im}(2\pi i S + \pi i T),$$

where S is the sum of the residues of the function $f : z \longrightarrow \dfrac{p(z)}{q(z)} e^{i\beta z}$ at those poles which lie in the upper half-plane, and T is the sum of the residues of f at the poles on the real axis.

(Note that by taking $\beta = 0$ we have an extension of Theorem 3).

Theorem 8(b)

Let p and q be polynomials with the following properties:

(i) q has simple zeros on the real axis;

(ii)   the degree of q exceeds that of p by at least one.

Then, if $\beta > 0$,

$$\int_{-\infty}^{\infty} \frac{p(t)}{q(t)} \cos \beta t \, dt = \mathrm{Re}(2\pi i S + \pi i T)$$

and

$$\int_{-\infty}^{\infty} \frac{p(t)}{q(t)} \sin \beta t \, dt = \mathrm{Im}(2\pi i S + \pi i T).$$

where S and T are as in Theorem 8(a).

(Note that Theorem 8(b) does not hold if $\beta = 0$.)

We shall leave the proofs of these theorems as an exercise and conclude this section with a couple of examples.

Example 4

Evaluate $\displaystyle\int_{-\infty}^{\infty} \frac{t}{t^3 - 1} \, dt$.

Solution

The poles of $z \longrightarrow \dfrac{z}{z^3 - 1}$ occur at 1 (with residue $\tfrac{1}{3}$), at $\tfrac{1}{2}(-1 + i\sqrt{3})$ (with residue $\tfrac{1}{6}(-1 - i\sqrt{3})$), and at $\tfrac{1}{2}(-1 - i\sqrt{3})$, which lies in the lower half-plane and so does not concern us.

Hence, by Theorem 8(a) with $\beta = 0$,

$$\int_{-\infty}^{\infty} \frac{t}{t^3 - 1} \, dt = \mathrm{Re}(2\pi i \cdot \tfrac{1}{6}(-1 - i\sqrt{3}) + \pi i \cdot \tfrac{1}{3})$$

$$= \frac{\pi\sqrt{3}}{3}.$$

Example 5

Evaluate $\displaystyle\int_{-\infty}^{\infty} \frac{\sin t}{t} \, dt$.

Solution

The function $z \longrightarrow \dfrac{e^{iz}}{z}$ has a simple pole at 0 with residue 1. Hence, by Theorem 8(b) with $\beta = 1$,

$$\int_{-\infty}^{\infty} \frac{\sin t}{t} \, dt = \mathrm{Im}(\pi i \cdot 1) = \pi.$$

Summary

In this section we have used a residue approach to integrate rational functions and functions of the form $t \longrightarrow \dfrac{p(t)}{q(t)} \cos \beta t$ and $t \longrightarrow \dfrac{p(t)}{q(t)} \sin \beta t$. In particular, we proved Jordan's Lemma, which is used to help with the estimation of an integral along a semicircle, and we showed how one can deal with simple poles lying on the real axis.

Although this section has been rather lengthy, we have really only scratched the surface of a large body of material. For example, we have been concerned only with a very small class of functions, and we have used only semicircular contours. In *Unit 11*, we shall see how the method described here can be applied to a wider class of functions, such as logarithm and power functions, and in the next problems section and in Section 10.6 we shall show how rectangular contours can be used to solve various types of problem.

Note You may find the following way of *writing* integrals along a closed semicircular contour and along the associated semicircle helpful:

$$\int_{\triangle} f(z)\,dz \ , \quad \int_{\frown} f(z)\,dz \ .$$

Such a way of writing contour integrals can obviously be extended to other types of contour. If you invent a notation which you wish to use in assessed work, be sure to define it the first time you use it.

Self-Assessment Questions

5. (i) State Jordan's Lemma.

   (ii) Why is it useful?

6. Use one of the theorems of this section to evaluate

   $$\int_{-\infty}^{\infty} \frac{\cos t}{t^2 + 1}\,dt \text{ and } \int_{-\infty}^{\infty} \frac{\sin t}{t^2 + 1}\,dt.$$

7. (i) Draw the contour you would use to evaluate

   $$\int_{-\infty}^{\infty} \frac{dt}{t(t-1)}.$$

   (ii) Use Theorem 8(a) to evaluate this integral.

8. Is the following statement true or is it false?

   If the function f has a pole on the real axis at a point α, then the value of $\int_{\Gamma} f$, where Γ is a semicircular (indentation) contour of radius ε, centre α (as illustrated in Fig. 9 on page 38), is $-\pi i \cdot \mathrm{Res}(f, \alpha)$.

Solutions

5. (i) See page 34.

   (ii) Jordan's Lemma is useful for estimating the integral of functions along the semicircle $\Gamma(r)$; in particular, it helps to deal with the case in which $f(z) = p(z)/q(z)$, where the degree of q is only one more than the degree of p.

6. By Theorem 7 (or Theorem 5) with $S = e^{-1}/2i$ (the residue of $z \longrightarrow e^{iz}/(z^2 + 1)$ at the point i), we get

$$\int_{-\infty}^{\infty} \frac{\cos t}{t^2 + 1}\, dt = \mathrm{Re}\!\left(2\pi i \cdot \frac{e^{-1}}{2i}\right) = \frac{\pi}{e},$$

and

$$\int_{-\infty}^{\infty} \frac{\sin t}{t^2 + 1}\, dt = \mathrm{Im}\!\left(2\pi i \cdot \frac{e^{-1}}{2i}\right) = 0.$$

(Note that the result of the second integral was to have been expected, since

$t \longrightarrow \dfrac{\sin t}{t^2 + 1}$ is an odd function.)

7. (i) The contour must avoid the poles at 0 and 1, so we can use the following one:

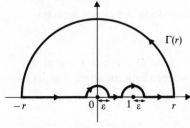

Fig. 10

   (ii) The residues at 0 and 1 are -1 and 1, respectively, so that, by Theorem 8(a) with $\beta = 0$, the value of the integral is

$$\mathrm{Re}(\pi i(-1 + 1)) = 0.$$

8. False. The value $-\pi i \cdot \mathrm{Res}(f, \alpha)$ is equal to $\displaystyle \lim_{\varepsilon \to 0^{+}} \int_{\Gamma} f.$

10.5 PROBLEMS

1. (This is a problem in real analysis: you may omit it if you are short of time.)

    (i) Prove that $\tan\theta > \theta$ if $0 < \theta < \pi/2$. (Hint: $\tan' = \sec^2$.)

    (ii) By considering the derivative of $h: \theta \longrightarrow (\sin\theta)/\theta$ and using part (i) prove that if $0 \leqslant \theta \leqslant \pi/2$, then $\sin\theta \geqslant 2\theta/\pi$. (This result was used in the proof of Jordan's Lemma.)

The following problems are designed to give you some practice in integrating along different types of contour. In general, we shall specify the contours involved unless the standard semicircular contour or a contour such as that in Fig. 9 is appropriate, in which case we say nothing.

2. Evaluate the integral

$$\int_0^\infty \frac{t\sin t}{t^4 + 4}\,dt,$$

by using the result of one of the theorems in the previous section.

(Hint: Why can you replace $\int_0^\infty$ by $\frac{1}{2}\int_{-\infty}^\infty$?)

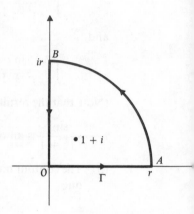

Fig. 11

3. Evaluate the integral in Problem 2 by integrating along the closed quarter-circle Γ shown in Fig. 11.

4. (a) Prove that, if the function f has a simple pole at the point α and if $\Gamma(\varepsilon)$ is part of a circular path of radius ε, subtending an angle θ at its centre α (see Fig. 12), then

$$\lim_{\varepsilon \to 0^+} \int_{\Gamma(\varepsilon)} f(z)\,dz = i\theta \cdot \operatorname{Res}(f, \alpha).$$

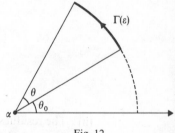

Fig. 12

    (b) Use the result of part (a) to prove Theorem 8(a). (Do not bother to give the full details—just outline the main steps of the argument. Theorem 8(b) can be proved in much the same way.)

5. By integrating the function $z \longrightarrow \dfrac{e^{\alpha z}}{e^z + 1}$ along a rectangular contour Γ with vertices r, $r + 2\pi i$, $-r + 2\pi i$, $-r$ (Fig. 13), show that, if $0 \leqslant \alpha \leqslant 1$, then

$$\int_{-\infty}^\infty \frac{e^{\alpha t}}{e^t + 1}\,dt = \frac{\pi}{\sin\pi\alpha}.$$

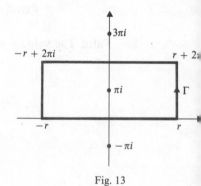

Fig. 13

Solutions

1. (i) Since $\tan = \sec^2$,

$$\tan \theta = \int_0^\theta \sec^2 \phi \, d\phi.$$

But $0 < \cos \phi < 1$ if $0 < \phi < \pi/2$, and so $\sec^2 \phi > 1$ if $0 < \phi < \pi/2$; therefore

$$\tan \theta \geqslant \int_0^\theta 1 \, d\phi = \theta, \quad \text{if } 0 < \theta < \pi/2.$$

(ii) First note that if $\theta = 0$ or $\theta = \pi/2$, then

$$\sin \theta = 2\theta/\pi.$$

The derivative of $h(\theta) = (\sin \theta)/\theta$ is

$$h'(\theta) = \frac{\theta - \tan \theta}{\theta^2 \cos \theta},$$

and so $h'(\theta) < 0$ if $0 < \theta < \pi/2$.

It follows that h is decreasing over the interval $[0, \pi/2]$, and so, if $0 \leqslant \theta \leqslant \pi/2$, then

$$\frac{\sin \theta}{\theta} \geqslant \frac{\sin \pi/2}{\pi/2} = \frac{2}{\pi}, \quad \text{as required.}$$

2. Since $t \longrightarrow \dfrac{t \sin t}{t^4 + 4}$ is an even function, we have

$$\int_0^\infty \frac{t \sin t}{t^4 + 4} \, dt = \frac{1}{2} \int_{-\infty}^\infty \frac{t \sin t}{t^4 + 4} \, dt.$$

The singularities of the function $z \longrightarrow \dfrac{z e^{iz}}{z^4 + 4}$ in the upper half-plane are simple poles at the points $\sqrt{2} e^{i\pi/4}$ and $\sqrt{2} e^{3i\pi/4}$ (that is, at $1 + i$ and $-1 + i$) with residues $\dfrac{e^{i(1+i)}}{8i}$ and $\dfrac{e^{i(-1+i)}}{-8i}$, respectively. Hence, applying Theorem 5 (or Theorem 7) with

$$S = \frac{e^{i(1+i)}}{8i} + \frac{e^{i(-1+i)}}{-8i} = \frac{e^{-1}}{8i}(e^i - e^{-i}) = \frac{1}{4e} \sin 1,$$

we get

$$\int_{-\infty}^\infty \frac{t \sin t}{t^4 + 4} \, dt = \operatorname{Im}\left(2\pi i \cdot \frac{1}{4e} \sin 1\right)$$

$$= \frac{\pi}{2e} \sin 1.$$

This gives

$$\int_0^\infty \frac{t \sin t}{t^4 + 4} \, dt = \frac{\pi}{4e} \sin 1.$$

3. Consider $\displaystyle\int_\Gamma \frac{z e^{iz}}{z^4 + 4} \, dz$, where Γ is the given contour (Fig. 11). The only pole of the integrand lying inside Γ is a simple pole at $1 + i$ with residue $\dfrac{e^{i(1+i)}}{8i}$.

Hence, by the Residue Theorem,

$$2\pi i \cdot \frac{1}{8i} e^{i(1+i)} = \int_\Gamma \frac{z e^{iz}}{z^4 + 4} \, dz$$

$$= \int_0^r \frac{t e^{it}}{t^4 + 4} \, dt + \int_{\Gamma(r)} \frac{z e^{iz}}{z^4 + 4} \, dz - \int_0^r \frac{(it) e^{i(it)} i}{(it)^4 + 4} \, dt,$$

where $\Gamma(r)$ is the quarter-circle from A to B (Fig. 11). (The last integral is obtained from the parametrization $\gamma(t) = it$, $0 \leqslant t \leqslant r$, for OB; the minus sign is included because we are integrating *from B to O*.)

Using the Estimation Theorem and Theorem 4 for the integral along $\Gamma(r)$ we have, for some constants K and r_0,

$$\left| \int_{\Gamma(r)} \frac{z e^{iz}}{z^4 + 4} \, dz \right| \leqslant \frac{K \cdot 1}{r^3} \cdot \frac{\pi r}{2} \,\, (\text{if } r \geqslant r_0), \quad \text{since } |e^{iz}| \leqslant 1.$$

It follows that the integral along $\Gamma(r)$ has limit zero as r becomes large. Hence

$$\frac{\pi}{4e} \cdot e^i = \int_0^\infty \frac{te^{it}}{t^4 + 4}\, dt + \int_0^\infty \frac{te^{-t}}{t^4 + 4}\, dt.$$

Equating imaginary parts, we get

$$\mathrm{Im}\left(\int_0^\infty \frac{te^{it}}{t^4 + 4}\, dt\right) = \int_0^\infty \frac{t \sin t}{t^4 + 4}\, dt = \mathrm{Im}\left(\frac{\pi}{4e} \cdot e^i\right) = \frac{\pi}{4e} \sin 1.$$

4. (a) Since f has a simple pole at the point α, we can write

$$f(z) = \frac{\mathrm{Res}(f, \alpha)}{z - \alpha} + g(z),$$

where g is analytic on some δ-neighbourhood of α.

It follows that, if $\varepsilon < \delta$, then

$$\int_{\Gamma(\varepsilon)} f(z)\, dz = \mathrm{Res}(f, \alpha) \int_{\Gamma(\varepsilon)} \frac{dz}{z - \alpha} + \int_{\Gamma(\varepsilon)} g(z)\, dz.$$

Since g is analytic on the disc $|z - \alpha| < \delta$, it is bounded on any closed disc $|z - \alpha| \leq \delta_1$ where $\varepsilon < \delta_1 < \delta$, and therefore on $\Gamma(\varepsilon)$ (by M, say), and so, by the Estimation Theorem

$$\left|\int_{\Gamma(\varepsilon)} g(z)\, dz\right| \leq M\varepsilon\theta,$$

which has limit zero as ε approaches zero.

Also,

$$\int_{\Gamma(\varepsilon)} \frac{dz}{z - \alpha} = \int_{\theta_0}^{\theta_0 + \theta} \frac{\varepsilon e^{it}}{\varepsilon e^{it}} i\, dt = i\theta,$$

using the parametrization $\gamma(t) = \alpha + \varepsilon e^{it}$, $\theta_0 \leq t \leq \theta_0 + \theta$.

Hence, $\int_{\Gamma(\varepsilon)} f(z)\, dz = i\theta \cdot \mathrm{Res}(f, \alpha)$.

(This result will be used in *Unit 14, Laplace Transforms*.)

(b) Let q have simple zeros at the points $c_1, \ldots, c_k$ on the real axis. Then if Γ is the closed contour shown in Fig. 14, it follows from the Residue Theorem that

$$2\pi i S = \int_\Gamma \frac{p(z)}{q(z)} e^{i\beta z}\, dz = \int_{\Gamma(r)} \frac{p(z)}{q(z)} e^{i\beta z}\, dz$$

$$+ \left(\int_{-r}^{c_1 - \varepsilon_1} + \int_{c_1 + \varepsilon_1}^{c_2 - \varepsilon_2} + \int_{c_k + \varepsilon_k}^{r}\right) \frac{p(t)}{q(t)} e^{i\beta t}\, dt$$

$$+ \left(\int_{\Gamma_1} + \int_{\Gamma_2} + \cdots + \int_{\Gamma_k}\right) \frac{p(z)}{q(z)} e^{i\beta z}\, dz,$$

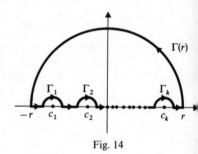

Fig. 14

where $\Gamma_1, \Gamma_2, \ldots, \Gamma_k$ are semicircular indentations at $c_1, c_2, \ldots, c_k$, of radii $\varepsilon_1, \varepsilon_2, \ldots, \varepsilon_k$, and $\Gamma(r)$ is the semicircle of radius r. As r becomes large, the first integral on the right has limit zero, as may be proved in the usual way. As each $\varepsilon_j, j = 1, 2, \ldots, k$, approaches zero, the corresponding integral $\int_{\Gamma_j} \frac{p(z)}{q(z)} e^{i\beta z}\, dz$ has limit $-i\pi \times$ the residue at c_j, by part (a). (The minus sign appears, since we are integrating *clockwise*.) Hence, letting r become large and letting each ε_j approach zero, we get

$$2\pi i S = \int_{-\infty}^\infty \frac{p(t)}{q(t)} e^{i\beta t}\, dt - i\pi T.$$

The result now follows immediately.

5. The function $z \longrightarrow \dfrac{e^{\alpha z}}{e^z + 1}$ has simple poles at the points $\pm\pi i, \pm 3\pi i, \pm 5\pi i, \ldots$; of these, only the point πi lies inside the contour Γ (Fig. 13), and the residue of the function there is $\dfrac{e^{\alpha\pi i}}{e^{\pi i}} = -e^{\alpha\pi i}$.

Hence, by the Residue Theorem, and using appropriate parametrizations,

$$2\pi i(-e^{\alpha\pi i}) = \int_{-r}^r \frac{e^{\alpha t}}{e^t + 1}\, dt + \int_0^{2\pi} \frac{e^{\alpha(r + it)}}{e^{r + it} + 1} i\, dt - \int_{-r}^r \frac{e^{\alpha(t + 2\pi i)}}{e^{t + 2\pi i} + 1}\, dt$$

$$- \int_0^{2\pi} \frac{e^{\alpha(-r + it)}}{e^{-r + it} + 1} i\, dt.$$

Since $e^{2\pi i} = 1$, the third integral on the right-hand side becomes $-e^{2\alpha\pi i}\int_{-r}^{r}\dfrac{e^{\alpha t}}{e^t + 1}\,dt$.

Hence,

$$-2\pi i e^{\alpha\pi i} = (1 - e^{2\alpha\pi i})\int_{-r}^{r}\frac{e^{\alpha t}}{e^t + 1}\,dt + ie^{\alpha r}\int_{0}^{2\pi}\frac{e^{\alpha it}}{e^{r + it} + 1}\,dt$$

$$- ie^{-\alpha r}\int_{0}^{2\pi}\frac{e^{\alpha it}}{e^{-r + it} + 1}\,dt.$$

We shall now show that the last two integrals have limit zero as r becomes large. This follows from the Estimation Theorem, since

$$\left| ie^{\alpha r}\int_{0}^{2\pi}\frac{e^{\alpha it}}{e^{r + it} + 1}\,dt \right| \leqslant e^{\alpha r}\cdot\frac{1}{e^r - 1}\cdot 2\pi,$$

and

$$\left| ie^{-\alpha r}\int_{0}^{2\pi}\frac{e^{\alpha it}}{e^{-r + it} + 1}\,dt \right| \leqslant e^{-\alpha r}\cdot\frac{1}{1 - e^{-r}}\cdot 2\pi,$$

and both of these have limit zero as r becomes large (α being such that $0 \leqslant \alpha \leqslant 1$).

Hence

$$\int_{-\infty}^{\infty}\frac{e^{\alpha t}}{e^t + 1}\,dt = \frac{-2\pi i e^{\alpha\pi i}}{1 - e^{2\alpha\pi i}} = \frac{2\pi i}{e^{\alpha\pi i} - e^{-\alpha\pi i}} = \frac{\pi}{\sin \pi\alpha}.$$

(Notice the reason for choosing a rectangular contour. If we had taken a semicircular one, then, on letting r become large, we should introduce more and more poles of the function inside the contour. The rectangular contour ensures that there is only one pole to deal with, whatever the value of r.)

10.6 THE SUMMATION OF SERIES

In this section, we shall show how residues can be used to sum certain series of the form

$$\sum_{n=1}^{\infty} \phi(n) \quad \text{and} \quad \sum_{n=1}^{\infty} (-1)^n \phi(n),$$

where ϕ is a suitable even function. In particular, we shall use residues to sum the series $\sum_{n=1}^{\infty} \dfrac{1}{n^2}$, $\sum_{n=1}^{\infty} \dfrac{(-1)^n}{\alpha^2 - n^2}$ and $\sum_{n=1}^{\infty} \dfrac{1}{n^4}$. (The corresponding functions ϕ are $z \longrightarrow \dfrac{1}{z^2}$, $z \longrightarrow \dfrac{1}{\alpha^2 - z^2}$ and $z \longrightarrow \dfrac{1}{z^4}$.)

At first sight, you might find it somewhat strange that the theory of residues (which is, after all, a method for evaluating integrals) should also be useful for summing series. However, if you consider that we can integrate a function along a contour Γ by taking

$$2\pi i \times \text{(the sum of its residues at points inside } \Gamma),$$

it seems at least possible that we can make each such residue a term of the series to be summed, so that we may evaluate the required sum by means of the integral of some related function.

In fact, this is quite easily done. All we shall need for this are two facts that were proved in Section 10.1 concerning the functions $z \longrightarrow \pi \cot \pi z \left(= \dfrac{\pi \cos \pi z}{\sin \pi z} \right)$ and $z \longrightarrow \pi \operatorname{cosec} \pi z \left(= \dfrac{\pi}{\sin \pi z} \right)$, namely:

(i) the function $z \longrightarrow \pi \cot \pi z$ has simple poles at the integers $0, \pm 1, \pm 2, \ldots$, and the residue at any of these poles is 1 (Self-Assessment Question 5(i) on page 15);

(ii) the function $z \longrightarrow \pi \operatorname{cosec} \pi z$ has simple poles at the integers $0, \pm 1, \pm 2, \ldots$, and the residue at the point n is $(-1)^n$ (Example 4 on page 10).

To see why these are relevant, let us first consider the function

$$f(z) = \phi(z)\pi \cot \pi z,$$

where ϕ is an even function with a finite number of poles, none of which occurs at an integer, except (possibly) at zero. Then the residue of f at a non-zero integer n is equal to $\phi(n)$, and the sum of the residues at the positive integers is equal to $\sum_{n=1}^{\infty} \phi(n)$, the sum we wish to evaluate. So our aim is (essentially) to integrate along a contour which includes the positive integers. The following theorem shows how this can be done.

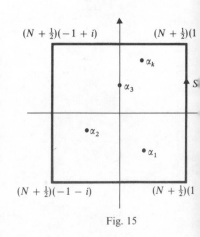

Fig. 15

Theorem 9

Let ϕ be an even function with poles at $\alpha_1, \ldots, \alpha_k$ (none of which is an integer) and possibly at 0, and let S_N be the square contour whose vertices are the points $(N + \frac{1}{2})(1 + i)$, $(N + \frac{1}{2})(-1 + i)$, $(N + \frac{1}{2})(-1 - i)$, $(N + \frac{1}{2})(1 - i)$, where N is a positive integer large enough for S_N to contain all of $\alpha_1, \ldots, \alpha_k$. Suppose, further that

$$\lim_{N \to \infty} \int_{S_N} f(z) \, dz = 0, \quad \text{where } f(z) = \phi(z)\pi \cot \pi z.$$

Then $\sum_{n=1}^{\infty} \phi(n)$ exists, and is equal to

$$-\frac{1}{2}\left(\operatorname{Res}(f, 0) + \sum_{r=1}^{k} \operatorname{Res}(f, \alpha_r) \right).$$

Proof

Let $f(z) = \phi(z)\pi \cot \pi z$. Then, as we remarked earlier, f has a simple pole at each non-zero integer n, and its residue there is $\phi(n)$. Hence, by the Residue Theorem,

$$\int_{S_N} \phi(z)\pi \cot \pi z \, dz$$

$$= 2\pi i \left(\sum_{n=1}^{N} \phi(n) + \sum_{n=1}^{N} \phi(-n) + \operatorname{Res}(f, 0) + \sum_{r=1}^{k} \operatorname{Res}(f, \alpha_r) \right)$$

$$= 2\pi i \left(2 \sum_{n=1}^{N} \phi(n) + \operatorname{Res}(f, 0) + \sum_{r=1}^{k} \operatorname{Res}(f, \alpha_r) \right),$$

since ϕ is an even function.

The result now follows by letting N become arbitrarily large, and using the fact that

$$\lim_{N \to \infty} \int_{S_N} \phi(z)\pi \cot \pi z \, dz = 0. \quad \blacksquare$$

The condition that $\displaystyle\lim_{N \to \infty} \int_{S_N} \phi(z)\pi \cot \pi z \, dz = 0$ is not nearly as artificial as it may seem at first sight. We shall see soon that $z \longrightarrow \cot \pi z$ is bounded on the squares S_N, and using this fact, it is then a simple matter to prove that the condition holds for a wide variety of functions ϕ. In fact, the reason we chose the square contour S_N rather than a circular contour of radius N is that the proof that $z \longrightarrow \cot \pi z$ is bounded is far simpler for S_N than the corresponding proof for circles. (If you are short of time, you should skip the proof, and assume the result.)

Theorem 10

There is a positive constant K such that, if S_N is the square contour whose vertices are the points $(N + \frac{1}{2})(1 + i)$, $(N + \frac{1}{2})(-1 + i)$, $(N + \frac{1}{2})(-1 - i)$, $(N + \frac{1}{2})(1 - i)$, where N is a positive integer, then

$$|\cot \pi z| \leqslant K, \quad \text{for all } z \text{ on } S_N.$$

(Note that K is independent of N.)

Proof

If $z = x + iy$, where $y \geqslant 1$, then

$$|\cot \pi z| = \left| \frac{\cos \pi z}{\sin \pi z} \right| = \left| i \left(\frac{e^{\pi i z} + e^{-\pi i z}}{e^{\pi i z} - e^{-\pi i z}} \right) \right|$$

$$= \left| \frac{e^{2\pi i z} + 1}{e^{2\pi i z} - 1} \right|$$

$$\leqslant \frac{1 + e^{-2\pi y}}{1 - e^{-2\pi y}}$$

$$\leqslant \frac{2}{1 - e^{-2\pi}}, \quad \text{since } y \geqslant 1.$$

The same bound also holds for $y \leqslant -1$, as you can easily show.

It follows that the function $z \longrightarrow \cot \pi z$ is bounded on the horizontal sides of S_N for each integer N. So we have only to show that $z \longrightarrow \cot \pi z$ is bounded on the vertical sides of S_N. We have already dealt with those parts of the vertical sides for which $|y| > 1$; also the function $z \longrightarrow \cot \pi z$ is a periodic function with period 1. So it is enough to prove that $z \longrightarrow \cot \pi z$ is bounded on the set $\{z : \operatorname{Re} z = \frac{1}{2}, -1 \leqslant \operatorname{Im} z \leqslant 1\}$. We can then deduce that $z \longrightarrow \cot \pi z$ is bounded on the sets $\{z : \operatorname{Re} z = N + \frac{1}{2}, -1 \leqslant \operatorname{Im} z \leqslant 1\}$ for each integer N, and the result then follows.

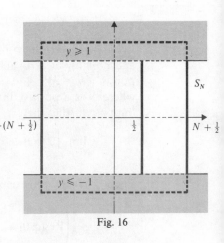

Fig. 16

47

But $z \longrightarrow \cot \pi z$ is continuous on the closed set $\{z : \operatorname{Re} z = \frac{1}{2}, -1 \leqslant \operatorname{Im} z \leqslant 1\}$, and so it must be bounded there, thus completing the proof. ∎

Let us now use the method of Theorem 9 to sum a particular series. The example we shall choose is one which you have probably met several times before.

Example 1

Prove that $\displaystyle\sum_{n=1}^{\infty} \frac{1}{n^2} = \frac{\pi^2}{6}$.

Solution

Let $\phi(z) = 1/z^2$ and $f(z) = (\pi \cot \pi z)/z^2$. Then f has a simple pole at each integer $n\,(\neq 0)$ with residue $1/n^2$, and a pole of order three at 0 with residue $-\pi^2/3$. There are no other poles.

Choose K as in Theorem 10. Then if z lies on the square contour S_N, we have $|z| \geqslant N + \frac{1}{2}$; so that, for some constant K,

$$\left| \frac{\pi \cot \pi z}{z^2} \right| \leqslant \frac{\pi K}{(N + \frac{1}{2})^2}, \quad \text{for } z \text{ on } S_N.$$

It follows that

$$\left| \int_{S_N} f(z)\, dz \right| \leqslant \frac{\pi K}{(N + \frac{1}{2})^2} \cdot 4(2N + 1) = \frac{16\pi K}{2N + 1},$$

which has limit zero as N becomes large.

We can therefore apply Theorem 9 to deduce that

$$\sum_{n=1}^{\infty} \frac{1}{n^2} = -\tfrac{1}{2}\operatorname{Res}(f, 0) = -\tfrac{1}{2} \cdot \left(\frac{-\pi^2}{3} \right) = \frac{\pi^2}{6}.$$

The method we have just described can be modified to sum various different types of series. For example, how do you think we should set about summing a series of the form

$$\sum_{n=1}^{\infty} (-1)^n \phi(n)?$$

If you look back at the reason for introducing $z \longrightarrow \pi \cot \pi z$, namely that it has simple poles at the integers, with residue 1, then it seems reasonable to assume that we shall need to introduce another function, which also has a simple pole at each integer n, but this time with residue $(-1)^n$. Such a function is $z \longrightarrow \pi \operatorname{cosec} \pi z$, as we recalled at the beginning of this section. The method works, just as before, using the easily-proved fact that $|\operatorname{cosec} \pi z| \leqslant K'$ for z on the square contour S_N, where K' is another constant. In this case, the basic theorem then takes the following form:

Theorem 11

Let ϕ be an even function with poles at $\alpha_1, \ldots, \alpha_k$ (none of which is an integer) and possibly at zero, and let S_N be the square contour whose vertices are the points $(N + \frac{1}{2})(1 + i)$, $(N + \frac{1}{2})(-1 + i)$, $(N + \frac{1}{2})(-1 - i)$, $(N + \frac{1}{2})(1 - i)$, where N is a positive integer large enough for S_N to contain all of $\alpha_1, \ldots, \alpha_k$. Suppose further that

$$\lim_{N \to \infty} \int_{S_N} g(z)\, dz = 0, \quad \text{where } g(z) = \phi(z)\pi \operatorname{cosec} \pi z.$$

Then $\displaystyle\sum_{n=1}^{\infty} (-1)^n \phi(n)$ exists, and is equal to

$$-\frac{1}{2}\left(\operatorname{Res}(g, 0) + \sum_{r=1}^{k} \operatorname{Res}(g, \alpha_r) \right).$$

The method of proof of this theorem is the same as for Theorem 9, so we shall omit it. (You can try it as an optional problem if you wish.)

Example 2

Evaluate $\sum_{n=1}^{\infty} \frac{(-1)^n}{\alpha^2 - n^2}$, where α is any complex number which is not an integer.

Solution

Let

$$\phi(z) = \frac{1}{\alpha^2 - z^2}, \text{ and } g(z) = \frac{\pi \cosec \pi z}{\alpha^2 - z^2}.$$

Then g has a simple pole at each integer n (including 0) with residue $\frac{(-1)^n}{\alpha^2 - n^2}$, and simple poles at α and $-\alpha$ each with residue $\frac{-\pi \cosec \pi \alpha}{2\alpha}$. There are no other poles.

On the square contour S_N (where $N > |\alpha|$), we have $|z| \geq N + \frac{1}{2}$, so that, for some constant K',

$$\left| \frac{\pi \cosec \pi z}{\alpha^2 - z^2} \right| \leq \frac{\pi K'}{(N + \frac{1}{2})^2 - |\alpha|^2} \quad \text{for } z \text{ on } S_N.$$

It follows that

$$\left| \int_{S_N} f(z)\, dz \right| \leq \frac{\pi K'}{(N + \frac{1}{2})^2 - |\alpha|^2} \cdot 4(2N + 1),$$

which has limit zero as N becomes large.

We can therefore apply Theorem 11 to deduce that

$$\sum_{n=1}^{\infty} \frac{(-1)^n}{\alpha^2 - n^2} = -\tfrac{1}{2}(\text{Res}(g, 0) + \text{Res}(g, \alpha) + \text{Res}(g, -\alpha))$$

$$= -\frac{1}{2}\left(\frac{1}{\alpha^2} - \frac{\pi \cosec \pi \alpha}{2\alpha} - \frac{\pi \cosec \pi \alpha}{2\alpha} \right)$$

$$= \frac{1}{2\alpha^2}(\pi \alpha \cosec \pi \alpha - 1).$$

Summary

In this section, we showed how the problem of summing a series can sometimes be replaced by one of evaluating a contour integral using the Residue Theorem. In particular, we showed how series of the form

$$\sum_{n=1}^{\infty} \phi(n) \text{ and } \sum_{n=1}^{\infty} (-1)^n \phi(n), \quad \text{where } \phi \text{ is an even function,}$$

can be summed using this technique.

Self-Assessment Questions

1. Why did we introduce the function $z \longrightarrow \pi \operatorname{cosec} \pi z$ in order to sum certain series?

2. Fill in the blanks in the solution to the following problem.

Problem   Sum the series $\displaystyle\sum_{n=1}^{\infty} \frac{(-1)^n}{n^2}$.

Solution   Consider the function $g(z) = $ | (i) |.

This function has a simple pole at each non-zero integer n with residue

| (ii) |, and a pole of order three at 0 with residue $\dfrac{\pi^2}{6}$. Now

let S_N be

| (iii) |.

Then

$$|\operatorname{cosec} \pi z| \leqslant \boxed{\text{(iv)}} \quad \text{for } z \text{ on } S_N,$$

so that

$$\left| \int_{S_N} g(z)\, dz \right| \leqslant \boxed{\text{(v)}},$$

which | (vi) | as N becomes large.

Hence, by Theorem 11,

$$\sum_{n=1}^{\infty} \frac{(-1)^n}{n^2} = \boxed{\text{(vii)}}.$$

Solutions

1. The poles of $z \longrightarrow \pi \operatorname{cosec} \pi z$ occur at the integers, and the residue at the point n is equal to $(-1)^n$. Consequently, the residue of $z \longrightarrow \phi(z)\pi \operatorname{cosec} \pi z$ at the point n is equal to $(-1)^n \phi(n)$, where ϕ is a function which does not have simularities at the integers.

2. (i) $\dfrac{\pi \operatorname{cosec} \pi z}{z^2}$.

   (ii) $\dfrac{(-1)^n}{n^2}$.

   (iii)   the square contour with vertices $(N + \frac{1}{2})(1 + i)$, $(N + \frac{1}{2})(-1 + i)$, $(N + \frac{1}{2})(-1 - i)$, $(N + \frac{1}{2})(1 - i)$.

   (iv) K' (some constant).

   (v) $\dfrac{\pi K'}{(N + \frac{1}{2})^2} \cdot 4(2N + 1) = \dfrac{16\pi K'}{2N + 1}$.

   (vi)   has limit zero.

   (vii) $-\frac{1}{2}\operatorname{Res}(g, 0) = \dfrac{-\pi^2}{12}$.

10.7 PROBLEMS

1. Find the sum of the series $\displaystyle\sum_{n=1}^{\infty} \frac{1}{n^4}$.

$\left(\text{Hint: The residue of the function } z \longrightarrow \dfrac{\pi \cot \pi z}{z^4} \text{ at } 0 \text{ is equal to } -\dfrac{\pi^4}{45}.\right)$

2. Use the result of Example 2 of Section 10.6 to evaluate the following sums:

(i) $\displaystyle\sum_{n=1}^{\infty} \frac{(-1)^n}{4n^2 - 1}$; (ii) $\displaystyle\sum_{n=1}^{\infty} \frac{(-1)^n}{n^2 + 1}$.

3. (a) Prove that, if α is not an integer, then

$$\sum_{n=1}^{\infty} \frac{1}{n^2 - \alpha^2} = \frac{1}{2\alpha^2}(1 - \pi\alpha \cot \pi\alpha).$$

(b) By writing

$$\frac{1}{n^4 - 4} = \frac{1}{4}\left(\frac{1}{n^2 - 2} - \frac{1}{n^2 + 2}\right),$$

find the sum of the series $\displaystyle\sum_{n=1}^{\infty} \frac{1}{n^4 - 4}$.

(c) Describe *briefly* how you could extend the method of part (b) to find the sum of the series $\displaystyle\sum_{n=1}^{\infty} \frac{1}{n^6 - 8}$.

Solutions

1. Let $\phi(z) = \dfrac{1}{z^4}$, and $f(z) = \dfrac{\pi \cot \pi z}{z^4}$. Then f has a simple pole at each integer $n (\neq 0)$ with residue $\dfrac{1}{n^4}$, and a pole of order five at 0 with residue $-\dfrac{\pi^4}{45}$ (given). There are no other poles.

For z on the square contour S_N, we have $|z| \geq N + \frac{1}{2}$, so that, for some constant K,

$$\left|\frac{\pi \cot \pi z}{z^4}\right| \leq \frac{\pi K}{(N + \frac{1}{2})^4} \text{ for } z \text{ on } S_N.$$

It follows that

$$\left|\int_{S_N} f(z)\,dz\right| \leq \frac{\pi K}{(N + \frac{1}{2})^4} \cdot 4(2N + 1) = \frac{64\pi K}{(2N + 1)^3},$$

which has limit zero as N becomes large.

We can therefore apply Theorem 9 to deduce that

$$\sum_{n=1}^{\infty} \frac{1}{n^4} = -\tfrac{1}{2}\mathrm{Res}(f, 0) = -\tfrac{1}{2}\left(-\frac{\pi^4}{45}\right) = \frac{\pi^4}{90}.$$

2. (i) $\displaystyle\sum_{n=1}^{\infty} \frac{(-1)^n}{4n^2 - 1} = -\frac{1}{4}\sum_{n=1}^{\infty} \frac{(-1)^n}{(\frac{1}{2})^2 - n^2}$

$= -\frac{1}{4} \cdot \frac{1}{2(\frac{1}{2})^2} \cdot \left(\frac{\pi}{2}\operatorname{cosec}\frac{\pi}{2} - 1\right),$ by Example 2 (with $\alpha = \frac{1}{2}$),

$= -\frac{1}{4}(\pi - 2).$

(ii) $\displaystyle\sum_{n=1}^{\infty} \frac{(-1)^n}{n^2 + 1} = -\sum_{n=1}^{\infty} \frac{(-1)^n}{i^2 - n^2}$

$= -\frac{1}{2i^2}(\pi i \operatorname{cosec} \pi i - 1),$ by Example 2 (with $\alpha = i$),

$= \frac{1}{2}\left(\frac{\pi}{\sinh \pi} - 1\right),$ since $\operatorname{cosec} \pi i = 1/\sin \pi i$ and $\sin \pi i = i \sinh \pi$.

3.   (a)   Let $\phi(z) = \dfrac{1}{z^2 - \alpha^2}$, and $f(z) = \dfrac{\pi \cot \pi z}{z^2 - \alpha^2}$. Then f has a simple pole at each integer n (including 0) with residue $\dfrac{1}{n^2 - \alpha^2}$, and simple poles at α and $-\alpha$, both with residue $\dfrac{\pi \cot \pi \alpha}{2\alpha}$.

For z on the square contour S_N, we have $|z| \geqslant N + \frac{1}{2}$, so that, for some constant K,

$$\left| \frac{\pi \cot \pi z}{z^2 - \alpha^2} \right| \leqslant \frac{\pi K}{(N + \frac{1}{2})^2 - |\alpha|^2} \text{ for } z \text{ on } S_N.$$

It follows that

$$\left| \int_{S_N} f(z)\, dz \right| \leqslant \frac{\pi K}{(N + \frac{1}{2})^2 - |\alpha|^2} \cdot 4(2N + 1),$$

which has limit zero as N becomes large.

We can therefore apply Theorem 9 to deduce that

$$\sum_{n=1}^{\infty} \frac{1}{n^2 - \alpha^2} = -\tfrac{1}{2}[\operatorname{Res}(f, 0) + \operatorname{Res}(f, \alpha) + \operatorname{Res}(f, -\alpha)]$$

$$= -\frac{1}{2}\left(-\frac{1}{\alpha^2} + \frac{\pi \cot \pi \alpha}{2\alpha} + \frac{\pi \cot \pi \alpha}{2\alpha} \right)$$

$$= \frac{1}{2\alpha^2}(1 - \pi \alpha \cot \pi \alpha).$$

  (b)   By part (a),

$$\sum_{n=1}^{\infty} \frac{1}{n^2 - 2} = \tfrac{1}{4}(1 - \pi\sqrt{2} \cot \pi\sqrt{2}), \quad \text{taking } \alpha = \sqrt{2},$$

and

$$\sum_{n=1}^{\infty} \frac{1}{n^2 + 2} = -\tfrac{1}{4}[1 - \pi i \sqrt{2} \cot(\pi i \sqrt{2})], \quad \text{taking } \alpha = i\sqrt{2},$$

$$= -\frac{1}{4}\left(1 - \pi\sqrt{2}\frac{\cosh \pi\sqrt{2}}{\sinh \pi\sqrt{2}} \right),$$

since $\cos(i\pi\sqrt{2}) = \cosh \pi\sqrt{2}$, and $\sin(i\pi\sqrt{2}) = i \sinh \pi\sqrt{2}$.

Hence

$$\sum_{n=1}^{\infty} \frac{1}{n^4 - 4} = \frac{1}{4}\left(\sum_{n=1}^{\infty} \frac{1}{n^2 - 2} - \sum_{n=1}^{\infty} \frac{1}{n^2 + 2} \right)$$

$$= \frac{1}{16}(1 - \pi\sqrt{2} \cot \pi\sqrt{2}) + \frac{1}{16}\left(1 - \pi\sqrt{2}\frac{\cosh \pi\sqrt{2}}{\sinh \pi\sqrt{2}} \right)$$

$$= \frac{1}{8} - \frac{\pi\sqrt{2}}{16}\left(\frac{\cos \pi\sqrt{2}}{\sin \pi\sqrt{2}} + \frac{\cosh \pi\sqrt{2}}{\sinh \pi\sqrt{2}} \right).$$

  (c)   If ω is a complex cube root of 1 (so that $\omega^3 = 1$), we can write

$$n^6 - 8 = (n^2 - 2)(n^2 - 2\omega)(n^2 - 2\omega^2),$$

so that, using partial fractions,

$$\frac{1}{n^6 - 8} = \frac{A}{n^2 - 2} + \frac{B}{n^2 - 2\omega} + \frac{C}{n^2 - 2\omega^2},$$

for some constants A, B, C, which can easily be calculated. Taking the result of part (a) with $\alpha = \sqrt{2}$, $\sqrt{2}\omega^2$ and $\sqrt{2}\omega$, we can sum the three separate series, and hence deduce the sum of the series required.

10.8 MEROMORPHIC FUNCTIONS

In this section we shall show how the theory of residues can be used to give us information about the number of zeros and poles of a function inside a given closed contour. The result we obtain is called the *Principle of the Argument*, and we shall also show how it can be restated in terms of winding numbers.

But, first, a preliminary problem for you to work through.

Preliminary Problem

Let f be the function defined by $f(z) = \dfrac{(z + 3)(z + 1)}{z^2}$. Find the zeros and poles of f, $1/f$, f' and f'/f.

Solution

Since $f(z) = \dfrac{(z + 3)(z + 1)}{z^2}$, we get

$$\frac{1}{f(z)} = \frac{z^2}{(z + 3)(z + 1)}, \quad f'(z) = \frac{-(4z + 6)}{z^3}, \quad \text{and} \quad \frac{f'(z)}{f(z)} = \frac{-4(z + 6)}{z(z + 3)(z + 1)}.$$

We therefore get the following table of zeros and poles, with the orders of the zeros and poles shown in brackets:

	f	$1/f$	f'	f'/f
Zeros	$-3(1), -1(1)$	$0(2)$	$-\frac{3}{2}(1)$	$-\frac{3}{2}(1)$
Poles	$0(2)$	$-3(1), -1(1)$	$0(3)$	$0(1), -3(1), -1(1)$

Note that the poles of f'/f all occur at points which are either zeros or poles of f.

We have a special name for functions such as the one in the problem, whose only singularities are poles.

Definition

> A function f is called **meromorphic on a bounded region R** if the only singularities of f lying in R are poles.

The observation we made at the end of the solution of the preliminary problem applies more generally to any meromorphic function. We shall state the result as a lemma, and ask you to prove it in the next problem section.

Lemma

If f is a meromorphic function on a bounded region R, then the poles of f'/f in R all occur at points which are either zeros or poles of f.

Using this lemma, we can now prove the Principle of the Argument. The reason for the name will become clear in the next unit.

Theorem 12 (The Principle of the Argument)

Let f be a function, meromorphic on a simply-connected bounded region R, and let Γ be a simple-closed contour contained in R. Assume further that f has no zeros or poles *on* Γ. If N is the number of zeros and P is the number of poles of f *inside* Γ, each counted according to its order, then

$$N - P = \frac{1}{2\pi i} \int_\Gamma \frac{f'}{f}.$$

Remark The phrase 'each counted according to its order' means that, for example, a pole of order three would be counted three times.

Proof

Let us look at the singularities of f'/f. By the lemma, these can occur only at the poles or the zeros of f. So we must calculate the residues of f'/f at these points.

Suppose, first, that f has a pole of order k inside Γ at α. Then

$$f(z) = \frac{g(z)}{(z - \alpha)^k},$$

where g is analytic and non-zero on some neighbourhood of α. Differentiating, we get

$$f'(z) = \frac{g'(z)}{(z - \alpha)^k} - \frac{kg(z)}{(z - \alpha)^{k+1}} \quad (z \neq \alpha),$$

so that

$$\frac{f'(z)}{f(z)} = \frac{g'(z)}{g(z)} - \frac{k}{z - \alpha}.$$

It follows that f'/f has a simple pole at α, since g'/g is analytic on a neighbourhood of α, and the residue of f'/f at α is $-k$.

Suppose now that f has a zero of order l inside Γ at β. Then $f(z) = (z - \beta)^l h(z)$, where h is analytic and non-zero on some neighbourhood of β, and we can go through a procedure almost identical to that described above, to deduce that the residue of f'/f at β is equal to l (see Self-Assessment Question 2, below).

It follows from the Residue Theorem that

$$\int_\Gamma \frac{f'}{f} = 2\pi i \times (\text{sum of the residues inside } \Gamma)$$

$$= 2\pi i(N - P), \quad \text{as required.} \quad \blacksquare$$

Example

Verify Theorem 12 for the function f defined in the preliminary problem, and the circle $\Gamma = \{z : |z| = 2\}$.

Solution

In this case $N = 1$ (since -3 lies outside Γ), and $P = 2$ (since 0 is a pole of order two), so that $N - P = -1$. Also,

$$\frac{1}{2\pi i} \int_\Gamma \frac{f'(z)}{f(z)} \, dz = \frac{1}{2\pi i} \int_\Gamma \frac{-(4z + 6)}{z(z + 3)(z + 1)} \, dz$$

$$= \text{Res}(f'/f, 0) + \text{Res}(f'/f, -1), \text{ by the Residue Theorem,}$$

$$= -2 + 1 = -1.$$

The theorem is therefore verified in this particular case.

We conclude this section by restating the Principle of the Argument in terms of winding numbers. You may recall from the previous unit, that if Γ is a closed contour and $0 \notin \Gamma$, then the *winding number* $\mathrm{Wnd}(\Gamma, 0)$ of Γ about the point 0 is defined by the formula

$$\mathrm{Wnd}(\Gamma, 0) = \frac{1}{2\pi i} \int_{\Gamma} \frac{dw}{w}.$$

If, now, f is a meromorphic function as described in Theorem 12, then the image contour $f(\Gamma)$ is a closed (but not necessarily simple-closed) contour, and we get

$$\mathrm{Wnd}(f(\Gamma), 0) = \frac{1}{2\pi i} \int_{f(\Gamma)} \frac{dw}{w}$$

$$= \frac{1}{2\pi i} \int_{\Gamma} \frac{f'(z)}{f(z)} \, dz,$$

using the substitution $w = f(z)$, $dw = f'(z) \, dz$.

We can therefore deduce the following corollary to Theorem 12.

Corollary

With the above notation

$$\mathrm{Wnd}(f(\Gamma), 0) = N - P.$$

In the next unit, we shall return to the Principle of the Argument, and will see, in particular, how it can be used to describe the change in the value of the argument as we go round a contour.

Summary

In this section we defined the term meromorphic function, we proved the Principle of the Argument, and restated it in terms of winding numbers.

Self-Assessment Questions

1. Verify the result of Theorem 12 when $f(z) = \dfrac{(z-1)^2}{z}$ and $\Gamma = \{z : |z| = 2\}$.

2. Show that, if f has a zero of order l at β, then the residue of f'/f at β is equal to l.

3. Verify the above corollary if $f : z \longrightarrow z + 1/z$ and

   (i) $\Gamma = \{z : |z| = r > 1\}$,

   (ii) $\Gamma = \{z : |z| = r < 1\}$.

Solutions

1. In this case, $N = 2$, $P = 1$, so $N - P = 1$.

    But $f'(z) = \dfrac{z^2 - 1}{z^2}$, so that

    $$\frac{1}{2\pi i} \int_\Gamma \frac{f'(z)}{f(z)}\, dz = \frac{1}{2\pi i} \int_\Gamma \frac{z + 1}{z(z - 1)}\, dz$$

    $$= 2 + (-1), \quad \text{by the Residue Theorem,}$$

    $$= 1, \quad \text{as required.}$$

2. If f has a zero of order l at the point β, then

    $$f(z) = (z - \beta)^l h(z),$$

    where h is non-zero and analytic on a neighbourhood of β. Differentiating, we get

    $$f'(z) = l(z - \beta)^{l-1} h(z) + (z - \beta)^l h'(z),$$

    so that

    $$\frac{f'(z)}{f(z)} = \frac{l}{z - \beta} + \frac{h'(z)}{h(z)}.$$

    It follows that f'/f has a simple pole at β, since h'/h is analytic on a neighbourhood of β; the residue of f'/f at β is l.

3. (i) On Γ, $z = re^{i\theta}$, $0 \leqslant \theta \leqslant 2\pi$, so that if $w = f(z)$ then

    $$w = re^{i\theta} + \frac{1}{r}e^{-i\theta} = \left(r + \frac{1}{r}\right)\cos\theta + i\left(r - \frac{1}{r}\right)\sin\theta.$$

    But $u = \left(r + \dfrac{1}{r}\right)\cos\theta$, $v = \left(r - \dfrac{1}{r}\right)\sin\theta$, $0 \leqslant \theta < 2\pi$, is a parametrization of an ellipse. The image $f(\Gamma)$ of Γ is therefore an ellipse, traversed once in the positive direction if $r > 1$, so that $\text{Wnd}(f(\Gamma), 0) = 1$.

    The function f has a pole at 0 and zeros at $\pm i$, all lying inside Γ, so that $N = 2$ and $P = 1$. Hence, in this case, $\text{Wnd}(f(\Gamma), 0) = N - P$, as expected.

    (ii) If $r < 1$ then $r - \dfrac{1}{r} < 0$, so that the above argument applies except that the ellipse is now traversed once in the negative direction, so that $\text{Wnd}(f(\Gamma), 0) = -1$. The points $\pm i$, do not lie inside Γ, so that $N = 0$. Once again, $\text{Wnd}(f(\Gamma), 0) = N - P$.

10.9 PROBLEMS

1. By writing $f(z) = \dfrac{(z - \alpha_1)^{l_1} \ldots (z - \alpha_n)^{l_n}}{(z - \beta_1)^{k_1} \ldots (z - \beta_m)^{k_m}} g(z)$, where the α_j and β_j are, respectively, the zeros and poles of f in R, prove the Lemma (page 53), and deduce that all the poles of f'/f are simple poles.

2. Show how Theorem 12 can be modified to give an expression for $N_a - P$, where N_a is the number of solutions of the equation $f(z) = a$, inside Γ.

3. This problem ties together the ideas of Sections 10.6 and 10.8.

   (a) Let f be a function, meromorphic on a simply-connected bounded region R, and let Γ be a simple-closed contour contained in R on which f is analytic and non-zero. Suppose, further, that all of the zeros $(\alpha_1, \alpha_2, \ldots)$ of f are simple zeros, and that all of the poles $(\beta_1, \beta_2, \ldots)$ of f are simple poles, and that only finitely many of these poles lie inside Γ. Show that, for any function ϕ, analytic on R,

   $$\frac{1}{2\pi i} \int_\Gamma \frac{f'(z)}{f(z)} \phi(z)\, dz = \sum_j \phi(\alpha_j) - \sum_j \phi(\beta_j),$$

   where the summations extend over those zeros and poles which lie inside Γ.

   (b) What happens in part (a) if $f(z) = \sin \pi z$? Does the result seem familiar?

Solutions

1. By imitating the proof of Theorem 12, we can prove that

   $$\frac{f'(z)}{f(z)} = \sum_{j=1}^{n} \frac{l_j}{z - \alpha_j} - \sum_{j=1}^{m} \frac{k_j}{z - \beta_j} + \frac{g'(z)}{g(z)},$$

   where g'/g is analytic and non-zero on R. It follows that the poles of f'/f occur at the points $\alpha_1, \ldots, \alpha_n, \beta_1, \ldots, \beta_m$, and that they are all simple poles.

2. Replace $f(z)$ in Theorem 12 by $f(z) - a$. Then

   $$\frac{1}{2\pi i} \int_\Gamma \frac{f'(z)}{f(z) - a}\, dz = N_a - P.$$

   You can also check that $N_a - P = \mathrm{Wnd}(f(\Gamma), a)$, the winding number of $f(\Gamma)$ about a.

3. (a) If α_j is a simple zero of f lying inside Γ, then α_j is a simple pole of $\dfrac{f'\phi}{f}$, by the result of Problem 1, and the residue there is $\phi(\alpha_j)$. Similarly, if β_j is a simple pole of f lying inside Γ, then (again by Problem 1) β_j is a simple pole of $\dfrac{f'\phi}{f}$ and the residue there is $-\phi(\beta_j)$. The result now follows from the Residue Theorem.

   (b) If $f(z) = \sin \pi z$, then f has no poles, and the zeros occur at the integers. Also

   $$\frac{f'(z)}{f(z)} = \frac{\pi \cos \pi z}{\sin \pi z} = \pi \cot \pi z.$$

   Hence

   $$\int_\Gamma \phi(z)\, \pi \cot \pi z\, dz = \sum_n \phi(n),$$

   where the summation extends over those integers which lie inside Γ.

   This is essentially the result we proved in Section 10.6, and shows that the summation of series and the Principle of the Argument are more closely related than you might at first have thought!

Unit 11   Analytic Functions

Conventions

Before working through this text make sure you have read *A Guide to the Course: Complex Analysis.*

References to units of other Open University courses in mathematics take the form:

Unit M100 13, Integration II.

The set book for the course M231, Analysis, is M. Spivak, *Calculus*, paperback edition (W. A. Benjamin/Addison-Wesley, 1973). This is referred to as:

Spivak.

Optional Material

This course has been designed so that it is possible to make minor changes to the content in the light of experience. You should therefore consult the supplementary material to discover which sections of this text are not part of the course in the current academic year.

11.0   INTRODUCTION

By now, you should be well aware that, if a function is analytic, then it has very special properties. It follows that if, in a given problem, we can show that the functions involved *are* analytic, then we have a powerful array of techniques at our disposal. The first part of this unit is concerned with developing methods for determining whether functions are indeed analytic. Such methods will require a discussion of uniform convergence.

Very often the solution of a given problem is most easily obtained as an infinite series or as an integral, and we shall need to show that such representations can give us functions which are analytic. Having done this, we may still be left with one or two difficulties, as the following example shows.

The function ϕ defined by the infinite series $\phi(z) = \sum_{n=0}^{\infty} z^n$ can easily be shown to be a solution of the differential equation $f'(z) = [f(z)]^2$, but the series is convergent only if $|z| < 1$. But what if we are interested in the value taken by this solution of the differential equation at some point outside the disc $|z| < 1$?

As it happens, the domain of ϕ can be extended if we notice that $\phi(z) = \dfrac{1}{1-z}$,

but in what sense are ϕ and $z \longrightarrow \dfrac{1}{1-z}$ the "same" function? In the second part of this unit we discuss the possibility of extending the domain of a given function by the *process of analytic continuation*. When this process is taken as far as possible, so that, in some sense, we get the "largest" domain for a given function, we obtain what we call a *complete analytic function*, which is a function which has a surface for its domain, the so-called *Riemann surface* (Fig. 1). We pay particular attention to the logarithm and power functions since these are the functions that arise most commonly in practice.

Fig. 1   A Riemann surface

Finally, we give you some recompense for your efforts to understand the concepts of "uniform convergence" and "complete analytic function" by describing some applications of the theory which we have developed. Some of these applications are relevant to the last three units of the course.

In order to appreciate the development in this unit you will need to be very familiar with certain material you have met before. The appropriate references are given on pages 62 and 87.

Television and Radio

In the sixth radio programme associated with the course we shall discuss the concept of uniform convergence of a sequence of functions.

In the eighth television programme we shall look at analytic continuation and Riemann surfaces.

11.1 UNIFORM CONVERGENCE

In this section we shall develop the theory of uniform convergence of sequences and series of complex functions. Since this topic is closely related to our previous study of sequences and series in *Unit 6*, *Taylor Series*, and also to the theory of uniform convergence of sequences and series of real functions, you may wish to find time to glance through *Unit 6* and Chapter 23 of **Spivak** before studying this unit. We begin with some preliminary problems in order to help you recall some of the relevant material. You are strongly advised to attempt them and to read through the solutions before proceeding.

Preliminary Problems

1. Define: the series $\sum\limits_{n=0}^{\infty} a_n$ *converges* to the sum s, where $a_0, a_1, a_2, \ldots$ are complex numbers.

2. Prove that an absolutely convergent series of complex numbers is convergent.

3. If f is a function such that the series $\sum\limits_{n=0}^{\infty} a_n z^n$ converges to $f(z)$ on $\{z : |z| < k\}$, what can you say about the series

   (i) $\sum\limits_{n=1}^{\infty} n a_n z^{n-1}$ and (ii) $\sum\limits_{n=0}^{\infty} \dfrac{a_n z^{n+1}}{n+1}$?

4. Define: the sequence of real functions $\{f_n\}$ *converges uniformly* to f on a set $A \subset \mathbf{R}$.

5. If the sequence of real functions $\{f_n\}$ converges uniformly to f on an interval $[a, b]$, what can you say about the sequence $\{f_n'\}$?

Solutions

1. The series $\sum\limits_{n=0}^{\infty} a_n$ converges to s if $\lim\limits_{n \to \infty} \left(\sum\limits_{k=0}^{n} a_k \right)$ exists and equals s.

2. See Problem 2(b), part (iii), of Section 6.2 of *Unit 6*.

3. (i) The series $\sum\limits_{n=1}^{\infty} n a_n z^{n-1}$ converges on $\{z : |z| < k\}$ to the sum $f'(z)$ (see Theorem 8 of *Unit 6*).

   (ii) The series $\phi(z) = \sum\limits_{n=0}^{\infty} \dfrac{a_n z^{n+1}}{n+1}$ converges on $\{z : |z| < k\}$ and, since Taylor series may be differentiated term by term, $\phi' = f$. The series therefore represents an antiderivative (or primitive) of f.

4. The sequence of real functions $\{f_n\}$ converges uniformly to f on the set A if, for every $\varepsilon > 0$, there is some N such that, for all x in A, if $n > N$, then $|f(x) - f_n(x)| < \varepsilon$.

5. We *cannot* be sure that the sequence $\{f_n'\}$ converges to f' on $[a, b]$, or even that it converges. For example, if $\{f_n\}$ is the sequence of real functions defined by

   $$f_n(x) = \frac{\sin nx}{n},$$

   then $\{f_n\}$ converges uniformly to 0 on $\mathbf{R}$; but $f_n'(x) = \cos nx$, and $\{f_n'(x)\}$ is divergent for some values of x (for example, $x = \pi$). For real functions we have the following result (**Spivak**, Theorem 23–3):

   Suppose that $\{f_n\}$ is a sequence of functions which are differentiable on $[a, b]$, and that $\{f_n\}$ converges (pointwise) to f. Suppose, moreover, that the sequence of derived functions $\{f'_n\}$ converges uniformly on $[a, b]$ to some continuous function g. Then f is differentiable and

$$f'(x) = \lim_{n \to \infty} f'_n(x).$$

If you answered this question incorrectly then you should read again Chapter 23 of **Spivak**, paying particular attention to pages 412–418.

Sequences of Complex Functions

Let f be a finite sum of functions, each analytic on a region R; then f is also analytic on R, and, moreover, the derivative of f can be found by differentiating each term and adding these derivatives. Since many functions can be represented conveniently by "infinite sums" (that is, infinite series), it is natural to ask whether the results of real analysis can be extended. For instance, can we show that under suitable conditions, the sum of a series of analytic functions is analytic, and that differentiation term by term is justified? We have already seen that differentiating a Taylor series term by term is valid (Theorem 8 of *Unit 6*), but in order to investigate more general series we need to introduce the notion of uniform convergence for sequences of complex functions.

An **infinite sequence** of complex functions is a function from the natural numbers to the set of all complex functions defined on a set, A say. We extend the convention adopted in *Unit 6* and use the notation $\{f_n\}$ for the sequence $n \longrightarrow f_n$.

Our intention is to investigate the convergence of sequences of complex functions, and we begin by introducing the following definition.

Definition

> The sequence of functions $\{f_n\}$ is said to **converge pointwise to** f **on a set** A if $\lim_{n \to \infty} f_n(z) = f(z)$ for each z in A.

Example 1

The sequence of functions $\left\{z \longrightarrow \dfrac{1}{z - n}\right\}$ converges pointwise to the zero function on $\mathbf{C} - \mathbf{N}$ since, for any $\varepsilon > 0$, we have

$$\left| \frac{1}{z - n} \right| < \frac{1}{n - |z|}, \quad \text{if } n > |z|,$$

$$< \varepsilon, \quad \text{if } n > \frac{1}{\varepsilon} + |z|.$$

It follows that $\lim_{n \to \infty} \dfrac{1}{z - n} = 0$ if $z \in \mathbf{C} - \mathbf{N}$.

In the above example we have shown that it is possible to find a number, N say, such that, for any $\varepsilon > 0$, we have $|1/(z - n)| < \varepsilon$ if $n > N$. We may, for example, choose N to be the smallest integer larger than $(1/\varepsilon) + |z|$. This number N clearly depends on the choice of ε, and we would expect that a small ε would necessitate the choice of a correspondingly large N. It is very important to notice that the number N may also be dependent on the choice of z.

It may occur to us to ask whether, for a given ε, the *same* number N (dependent on ε) will serve for *all* z in the set $\mathbf{C} - \mathbf{N}$, and this leads us to our next example.

Example 2

Is it possible to find a natural number N such that $\left| \dfrac{1}{z - n} \right| < \dfrac{1}{10}$ if $n > N$ for *all* $z \in \mathbf{C} - \mathbf{N}$?

It is, in fact, impossible to find such a number N, and intuitively it is not difficult to see why this is the case. We have eliminated the natural numbers from $\mathbf{C}$ in order that the functions $z \longrightarrow \dfrac{1}{z-n}$ shall be well defined. This will not stop z from getting very close to a natural number, and so we can make $\left|\dfrac{1}{z-n}\right|$ very large. More formally, suppose that there were such a number N; then

$$\frac{1}{|z-(N+1)|} < \frac{1}{10} \quad \text{for all } z \in \mathbf{C} - \mathbf{N}.$$

However, the set $\mathbf{C} - \mathbf{N}$ contains points z for which $|z - (N+1)| < 10$, and so we have a contradiction.

Definition

> Let $\{f_n\}$ be a sequence of functions defined on a set A, and let f be a function which is also defined on A. Then f is called the **uniform limit of** $\{f_n\}$ **on** A if, for every $\varepsilon > 0$, there is some N (independent of z) such that, for all z in A,
>
> if $n > N$, then $|f(z) - f_n(z)| < \varepsilon$.
>
> We also say that $\{f_n\}$ **converges uniformly to** f **on** A.

You should notice particularly that the choice of N depends upon the choice of ε, but that it is independent of any particular value of z in A.

It is clear that a sequence $\{f_n\}$ which converges uniformly to f on a set A is also pointwise convergent to f on A, but our previous examples illustrate the fact that the converse does not hold.

Example 3

The sequence of functions $\left\{z \longrightarrow \dfrac{1}{z-n}\right\}$ converges uniformly to the zero function on the disc $D = \{z : |z| < 1\}$, since, given any $\varepsilon > 0$, we have

$$\frac{1}{|z-n|} < \frac{1}{n-1}, \quad \text{if } n > 1,$$

$$< \varepsilon, \quad \text{if } n > \frac{1}{\varepsilon} + 1.$$

We can therefore choose N to be the smallest natural number greater than $(1/\varepsilon) + 1$. Notice that N depends on ε, but not on z.

We are now in a position to prove various results for uniformly convergent sequences of complex functions, and we begin with two theorems which are almost identical to results which you have met for real functions (Theorems 23–1 and 23–2 of **Spivak**).

Theorem 1

Suppose that $\{f_n\}$ is a sequence of functions which can be integrated along a contour Γ, and that $\{f_n\}$ converges uniformly on Γ to a function f which can be integrated along Γ. Then

$$\int_\Gamma f = \lim_{n \to \infty} \int_\Gamma f_n.$$

Proof

Let $\varepsilon > 0$. There is some N such that for all $n > N$ we have

$$|f(z) - f_n(z)| < \varepsilon \quad \text{for all } z \in \Gamma.$$

Thus, if $n > N$ and L is the length of Γ, we have

$$\left| \int_\Gamma f(z)\, dz - \int_\Gamma f_n(z)\, dz \right| = \left| \int_\Gamma (f(z) - f_n(z))\, dz \right|$$

$$\leqslant \varepsilon L, \quad \text{by the Estimation Theorem.}$$

Since this is true for any $\varepsilon > 0$, it follows that

$$\int_\Gamma f = \lim_{n \to \infty} \int_\Gamma f_n. \quad \blacksquare$$

Theorem 2

Suppose that $\{f_n\}$ is a sequence of functions which are continuous on a set A and that $\{f_n\}$ converges uniformly to f on A. Then f is also continuous on A.

The proof is almost identical to that for the similar result for real functions, and we therefore leave it for you to try. (See Problem 4 of Section 11.2.)

We have now come to the point where the treatments for real and complex functions differ: we prove a remarkable result which has no analogue in real analysis. This result arises, like many others, from our strong definition of differentiability.

Theorem 3

Let R be a region and let $\{f_n\}$ be a sequence of functions each analytic on R. If $\{f_n\}$ converges uniformly to f on every closed disc contained in R, then (i) f is analytic on R; furthermore, (ii) $\{f_n'\}$ converges uniformly to f' on every closed disc in R.

Proof

(i) Let $z_0 \in R$, let $\{z : |z - z_0| \leqslant r\}$ be a closed disc in R, and let $D = \{z : |z - z_0| < r\}$. Since $\{f_n\}$ converges uniformly to f on $\{z : |z - z_0| \leqslant r\}$, it is clear that the convergence is also uniform on D. We wish to show that f is analytic on D, and to do this we use Morera's Theorem (*Unit 5, Cauchy's Theorem I*). Let Δ be any triangle in D (Fig. 2); then

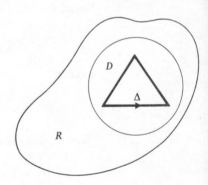

Fig. 2

$$\int_\Delta f = \lim_{n \to \infty} \int_\Delta f_n = 0,$$

where the first equality follows from Theorem 1 and the second from Cauchy's Theorem. By Morera's Theorem, f must be analytic on D, and hence analytic on R.

(ii) To show that $\{f_n'\}$ converges uniformly to f' on any closed disc in R we use Cauchy's Formula. Let $E = \{z : |z - z_0| \leqslant r\}$ be any closed disc in R. Since R is open we may choose $\rho > r$ so that $\{z : |z - z_0| \leqslant \rho\}$ is also in R. Let $C = \{z : |z - z_0| = \rho\}$ (Fig. 3). Then, for any $z \in E$, we have

$$f_n'(z) = \frac{1}{2\pi i} \int_C \frac{f_n(w)}{(w - z)^2}\, dw, \quad n = 1, 2, 3, \dots,$$

and

$$f'(z) = \frac{1}{2\pi i} \int_C \frac{f(w)}{(w - z)^2}\, dw.$$

We know that $\{f_n\}$ converges uniformly to f on the closed disc $\{z : |z - z_0| \leqslant \rho\}$, which lies in R. Thus, given $\varepsilon > 0$, we choose N such that

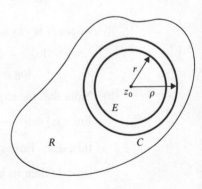

Fig. 3

$n > N$ implies that $|f(z) - f_n(z)| < \varepsilon$ for all z in this closed disc. In particular, since C is the boundary of this disc, $n > N$ implies that

$$|f(w) - f_n(w)| < \varepsilon \quad \text{for all } w \in C.$$

We now notice that, for any $z \in E$, we have

$$|f'(z) - f_n'(z)| = \left| \frac{1}{2\pi i} \int_C \frac{f(w) - f_n(w)}{(w - z)^2} \, dw \right|,$$

and we wish to estimate the modulus of the integral on the right. To do this we note that $|w - z| \geqslant \rho - r$ for any choice of $w \in C$ and $z \in E$. Hence, $n > N$ implies that

$$|f'(z) - f_n'(z)| \leqslant \frac{1}{2\pi} \cdot \frac{\varepsilon}{(\rho - r)^2} \cdot 2\pi\rho.$$

Since ρ and r are fixed numbers independent of the choice of $z \in E$ we have proved the required result. ∎

The above theorems enable us to integrate or differentiate a uniformly convergent sequence of analytic functions; this is in marked contrast to real analysis where differentiation is usually justified only if the resulting sequence is uniformly convergent.

Example 4

(a) Show that the sequence $\{s_n\}$, given by

$$s_n(z) = 1 + z + z^2 + \cdots + z^n,$$

converges uniformly to $z \longrightarrow \dfrac{1}{1 - z}$ on the disc $|z| \leqslant \rho$, for each $\rho < 1$.

(b) Show that $\{s_n\}$ is not uniformly convergent on the disc $|z| < 1$.

Solution

(a) Recall that the sum of a finite geometric series is

$$1 + z + z^2 + \cdots + z^n = \frac{1 - z^{n+1}}{1 - z}, \quad \text{if } z \neq 1,$$

so that

$$\left| (1 + z + z^2 + \cdots + z^n) - \frac{1}{1 - z} \right| = \frac{|z|^{n+1}}{|1 - z|}.$$

Now consider some fixed value of $\rho < 1$. If $|z| \leqslant \rho$, then

$$\left| (1 + z + z^2 + \cdots + z^n) - \frac{1}{1 - z} \right| = \frac{|z|^{n+1}}{|1 - z|} \leqslant \frac{\rho^{n+1}}{1 - \rho}.$$

Now let ε be an arbitrary positive number; then

$$\left| (1 + z + z^2 + \cdots + z^n) - \frac{1}{1 - z} \right| \leqslant \frac{\rho^{n+1}}{1 - \rho} < \varepsilon \quad \text{if } n > N,$$

where N is any natural number greater than

$$\frac{\log(\varepsilon - \varepsilon\rho)}{\log \rho}.$$

(b) Note that the sequence $\{s_n\}$ converges *pointwise* on the disc $|z| < 1$; in fact,

$\lim\limits_{n \to \infty} s_n(z) = \dfrac{1}{1 - z}$ if $|z| < 1$, but the sequence *is not uniformly convergent* on

this disc. For, suppose that $\{s_n\}$ is uniformly convergent on the disc $|z| < 1$; then its uniform limit must certainly be the function $z \longrightarrow \dfrac{1}{1 - z}$.

It follows that, for any $\varepsilon > 0$, there is a number N (independent of z) such that

$$\left| (1 + z + z^2 + \cdots + z^N) - \frac{1}{1 - z} \right| < \varepsilon$$

for all z in the disc $|z| < 1$. But

$$\left| (1 + z + z^2 + \cdots + z^N) - \frac{1}{1 - z} \right| = \frac{|z|^{N+1}}{|1 - z|},$$

and $\lim\limits_{\rho \to 1^-} \dfrac{\rho^{N+1}}{1 - \rho} = \infty$, so that there are points in the disc $|z| < 1$ for which $\dfrac{|z|^{N+1}}{|1 - z|}$ is as great as we please (and certainly not less than ε), and so we have a contradiction.

The above example provides us with an opportunity to verify the results of Theorems 2 and 3. Theorem 2 tells us that the function $z \longrightarrow \dfrac{1}{1 - z}$ is continuous on every closed disc $|z| \leqslant \rho$, where $\rho < 1$, which is indeed the case; Theorem 3 tells us that the sequence of functions $\{z \longrightarrow (1 + 2z + 3z^2 + \cdots + nz^{n-1})\}$ converges uniformly to $z \longrightarrow \dfrac{1}{(1 - z)^2}$ on every closed disc in the disc $|z| < 1$. We ask you to verify the second of these results in the first problems section.

Self-Assessment Questions

1. State three important properties of a sequence of functions $\{s_n\}$ each of which is analytic on a region R and uniformly convergent to s on every closed disc contained in R.

2. Show that for any $\rho > 0$ the sequence of functions $\left\{ z \longrightarrow \dfrac{1}{z - n} \right\}$ converges uniformly to the zero function on the disc $\{z : |z| \leqslant \rho\}$ with the natural numbers removed.

3. In the proof of Theorem 3 we have:

   "Let Δ be any triangle in D then $\displaystyle\int_\Delta f = \lim_{n \to \infty} \int_\Delta f_n = 0$."

   How do we know that $\displaystyle\int_\Delta f$ exists?

Solutions

1. (a) If Γ is any contour lying in a closed disc in R, then

    $$\int_\Gamma s = \lim_{n \to \infty} \int_\Gamma s_n \quad \text{(from Theorem 1).}$$

    (b) The function s is continuous on R (from Theorem 2).

    (c) The function s is analytic on R and $\{s_n'\}$ converges uniformly to s on every closed disc in R (from Theorem 3).

2. Given $\varepsilon > 0$, we have

    $$\frac{1}{|z - n|} < \frac{1}{n - \rho}, \quad \text{if } n > \rho \text{ (since } |z| \leqslant \rho),$$

    $$< \varepsilon, \quad \text{if } n > \frac{1}{\varepsilon} + \rho.$$

    We can therefore choose N to be any natural number greater than $(1/\varepsilon) + \rho$.

3. The functions f_n are analytic, and therefore continuous, on D. It follows from Theorem 2 that f is continuous on D, and, since Δ lies in D, the integral exists.

The Weierstrass M-test

From your study of real analysis you may recall that infinite sequences arise most often in the form of infinite series, and you may not be surprised to learn that the same is also true of complex analysis. You may also remember that the Weierstrass M-test provides a useful test for the uniform convergence of series of real functions, and, if you can bring to mind the proof of this result (**Spivak**, page 420), you will probably realize that it can easily be adapted to complex functions. First we require the following definition.

Definition

> The sequence of functions $\{f_n\}$, each defined on a set A, is said to be **uniformly summable to f on A** if the sequence $\{f_1 + f_2 + \cdots + f_n\}$ converges uniformly to f on A. We say that the series of functions $\sum\limits_{n=1}^{\infty} f_n$ **converges uniformly to f on A**, and write $f = \sum\limits_{n=1}^{\infty} f_n$.

Theorem 4 (The Weierstrass M-test)

Let $\{f_n\}$ be a sequence of functions defined on a set A, and suppose that $\{M_n\}$ is a sequence of real numbers such that $|f_n(z)| \leqslant M_n$ for all z in A. Suppose, moreover, that $\sum\limits_{n=1}^{\infty} M_n$ converges. Then, for each z in A, the series $\sum\limits_{n=1}^{\infty} f_n(z)$ converges (in fact, it converges absolutely), and $\sum\limits_{n=1}^{\infty} f_n$ converges uniformly on A to the function $z \longrightarrow \sum\limits_{n=1}^{\infty} f_n(z)$.

Proof

For each z in A the series $\sum\limits_{n=1}^{\infty} |f_n(z)|$ converges, by the comparison test, and consequently, from Theorem 3 of *Unit 6*, $\sum\limits_{n=1}^{\infty} f_n(z)$ converges (absolutely) to $f(z)$, say. Moreover, for all z in A, we have

$$
|f(z) - (f_1(z) + f_2(z) + \cdots + f_N(z))| = \left| \sum_{n=N+1}^{\infty} f_n(z) \right|
$$

$$
\leqslant \sum_{n=N+1}^{\infty} |f_n(z)|
$$

$$
\leqslant \sum_{n=N+1}^{\infty} M_n.
$$

Since $\sum\limits_{n=1}^{\infty} M_n$ converges, the number $\sum\limits_{n=N+1}^{\infty} M_n$ can be made as small as we please by choosing N (independent of z) sufficiently large, and so $\sum\limits_{n=1}^{\infty} f_n$ converges uniformly on A to $z \longrightarrow \sum\limits_{n=1}^{\infty} f_n(z)$. ∎

Applications of the Weierstrass M-test

As our first application of the Weierstrass M-test, we shall examine the convergence of power series.

Example 5

If f is a function analytic on the disc $|z| < \rho$ and $f(z) = \sum\limits_{n=0}^{\infty} a_n z^n$, show that, for each $r < \rho$, the series converges uniformly to f on the disc $|z| \leqslant r$.

Solution

Choose a number r_1 such that $r < r_1 < \rho$; then, from the corollary to Theorem 9 of *Unit 6* and Cauchy's formula for derivatives, we know that if $C = \{z : |z| = r_1\}$, then

$$a_n = \frac{1}{2\pi i} \int_C \frac{f(z)}{z^{n+1}} \, dz, \quad n = 1, 2, 3, \ldots.$$

But $|f|$ is bounded on the compact set C (by $M(r_1)$, say), so that

$$|a_n| \leqslant \frac{M(r_1)}{r_1^n} \quad \text{(see Problem 6(i) of Section 6.6 of \textit{Unit 6}).}$$

It follows that

$$|a_n z^n| \leqslant M(r_1) \left(\frac{r}{r_1}\right)^n, \quad \text{if } |z| \leqslant r;$$

but $\sum\limits_{n=0}^{\infty} \left(\frac{r}{r_1}\right)^n$ is convergent (since $r < r_1$), and the required result follows from the Weierstrass M-test.

An essential stage in the solution of many problems is to show that a certain function is analytic, and the Weierstrass M-test can be very useful for this purpose. As an illustration, consider the following example.

Example 6

If f is a function continuous on the interval $[0, 1]$, show that the function

$$\phi : z \longrightarrow \int_0^1 e^{zt} f(t) \, dt$$

is entire, and show that

$$\phi'(z) = \int_0^1 t e^{zt} f(t) \, dt.$$

Solution

For any fixed z the sequence of functions

$$\left\{ f_n : t \longrightarrow 1 + zt + \frac{z^2 t^2}{2!} + \cdots + \frac{z^n t^n}{n!} \right\}$$

is uniformly convergent to $t \longrightarrow e^{zt}$ on the interval $[0, 1]$, so that we may multiply f_n by f (see Problem 5 of Section 11.2), and integrate term by term (using Theorem 1), to obtain

$$\phi(z) = \sum_{n=0}^{\infty} a_n z^n,$$

where

$$a_n = \frac{1}{n!} \int_0^1 t^n f(t) \, dt.$$

But f is continuous on $[0, 1]$ and therefore $|f|$ is bounded (by M say) on the compact set $[0, 1]$, so that $|a_n| \leqslant \dfrac{M}{n!}$.

The Taylor series for ϕ at 0 therefore converges for all z, and thus ϕ is entire.

This example illustrates a common feature of such problems. We have what is perhaps a convenient representation of the function ϕ as an integral, but in order to establish certain properties of the function (in this case that it is entire) we require an alternative representation (in this case as a Taylor series).

For the remaining part we might set aside our scruples for one moment, and expect to be able to "differentiate under the integral sign" to obtain the required formula $\phi'(z) = \displaystyle\int_0^1 t e^{zt} f(t)\, dt$, but we have not yet proved that such an operation is valid.

In fact, we can establish the formula as follows. We have

$$\phi(z) = \sum_{n=0}^\infty a_n z^n \quad \text{for all } z \text{ in } \mathbf{C},$$

so that

$$\phi'(z) = \sum_{n=1}^\infty n a_n z^{n-1}, \quad \text{from Theorem 8 of } Unit\ 6,$$

$$= \sum_{n=1}^\infty \left(\int_0^1 \frac{t^n f(t)}{(n-1)!}\, dt \right) z^{n-1}$$

$$= \int_0^1 \left(\sum_{n=1}^\infty \frac{z^{n-1} t^n f(t)}{(n-1)!} \right) dt,$$

from Theorem 1 and the fact that the Taylor series
$$\sum_{n=1}^\infty \frac{t^n}{(n-1)!} \text{ is uniformly convergent on } [0, 1],$$

$$= \int_0^1 t e^{zt} f(t)\, dt.$$

This might lead you to suspect that "differentiation under the integral sign" can be justified in certain cases, and that it would be a useful technique to have at our disposal. We shall shortly prove a relevant result, but before doing so we look at some further examples which use the Weierstrass M-test directly.

Example 7

Find a region on which the function F defined by the series $F(z) = \displaystyle\sum_{n=1}^\infty \frac{\cos nz}{2^n}$ is analytic.

Solution

We notice that

$$F(z) = \sum_{n=1}^\infty \frac{\cos nz}{2^n} = \sum_{n=1}^\infty \frac{e^{inz} + e^{-inz}}{2^{n+1}}$$

is the sum of two series:

$$\phi_1(z) = \sum_{n=1}^\infty \frac{e^{inz}}{2^{n+1}} \quad \text{and} \quad \phi_2(z) = \sum_{n=1}^\infty \frac{e^{-inz}}{2^{n+1}}.$$

Let us concentrate our attention for the moment on the former series, and notice that, with $z = x + iy$,

$$\left|\frac{e^{inz}}{2^{n+1}}\right| = \frac{e^{-ny}}{2^{n+1}} = \frac{1}{2}\left(\frac{e^{-y}}{2}\right)^n.$$

But, if $y \geqslant -k$, we have $e^{-y} \leqslant e^k$, so that, if we choose $k < \log 2$, we have

$$\frac{e^{-y}}{2} \leqslant \frac{e^k}{2} < \frac{e^{\log 2}}{2} = 1.$$

The series $\sum_{n=1}^{\infty} \frac{1}{2}\left(\frac{e^k}{2}\right)^n$ converges and therefore the series $\sum_{n=1}^{\infty} \frac{e^{inz}}{2^{n+1}}$ converges uniformly on the half-plane $\{z : \operatorname{Im} z \geqslant -k\}$ for every $k < \log 2$, and hence on every closed disc in the half-plane $\{z : \operatorname{Im} z > -\log 2\}$. It follows from Theorem 3 that ϕ_1 is analytic on this half-plane.

The function ϕ_1 is undefined if $\operatorname{Im} z \leqslant -\log 2$, for in this case the series $\sum_{n=1}^{\infty} \frac{e^{inz}}{2^{n+1}}$ is clearly divergent (because the terms do not converge to zero).

In a similar fashion, we can show that the function ϕ_2 defined by the series $\phi_2(z) = \sum_{n=1}^{\infty} \frac{e^{-inz}}{2^{n+1}}$ is analytic if $\operatorname{Im} z < \log 2$ and undefined otherwise.

Since $F = \phi_1 + \phi_2$, we deduce that F is analytic on the infinite strip $\{z : -\log 2 < \operatorname{Im} z < \log 2\}$ and undefined elsewhere (Fig. 4).

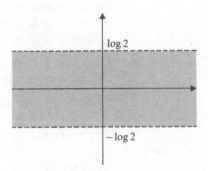

Fig. 4

Self-Assessment Questions

4. State the Weierstrass M-test.

5. Show that the series $\sum_{n=1}^{\infty} \frac{1}{n^2 + |z|^2}$ converges uniformly on $\mathbf{C}$.

6. Show that $\zeta(z) = \sum_{n=1}^{\infty} \frac{1}{n^z}$ defines an analytic function on $\{z : \operatorname{Re} z > 1\}$. (This function is called Riemann's zeta function.)

Solutions

4. See page 68.

5. Note that $\frac{1}{n^2 + |z|^2} \leqslant \frac{1}{n^2}$ and that $\sum_{n=1}^{\infty} \frac{1}{n^2}$ converges. The required result follows from the Weierstrass M-test.

6. We have

$$|n^{-z}| = |e^{-z \operatorname{Log} n}| = |e^{-\operatorname{Re} z \operatorname{Log} n} \cdot e^{-i \cdot \operatorname{Im} z \operatorname{Log} n}|$$

$$= e^{-\operatorname{Re} z \operatorname{Log} n} \cdot 1 = n^{-\operatorname{Re} z}.$$

So $|n^{-z}| \leqslant n^{-p}$, if $p > 1 \geqslant \operatorname{Re} z$, and so by the Weierstrass M-test the series converges uniformly (because $\sum_{n=1}^{\infty} n^{-p}$ converges if $p > 1$—see page 106 of *Unit 6*). Hence, by Theorem 3, the function $\zeta(z)$ is analytic on $\{z : \operatorname{Re} z > 1\}$.

Summary

In this section we have developed a method for testing functions defined by infinite series for analyticity: it relies upon the concept of uniform convergence. We also developed the Weierstrass M-test which enables us to determine whether certain types of series are uniformly convergent.

71

11.2 PROBLEMS

1. Verify that the sequence of functions $\{z \longrightarrow (1 + 2z + 3z^2 + \cdots + nz^{n-1})\}$ converges uniformly to $z \longrightarrow \dfrac{1}{(1-z)^2}$ on every closed disc in the disc $|z| < 1$.

2. Show that for each $a > 0$ the sequence of functions $\left\{z \longrightarrow \dfrac{\sin z}{z - (a+n)}\right\}$ converges uniformly to the zero function on the disc $|z| \leqslant a$.

3. Find a region on which the series $f(z) = \sum\limits_{n=0}^{\infty} e^{nz}$ defines an analytic function f.

4. Prove Theorem 2.

5. If the sequence of functions $\{f_n\}$ converges uniformly to f on a compact set A and g is a function continuous on A, show that $\{gf_n\}$ converges uniformly to gf on A.

6. Show that for any $\rho < 1$ the sequence of functions $\{z \longrightarrow z^n + 3z - 1\}$ converges uniformly on the disc $|z| \leqslant \rho$.

7. Show that the function f defined by the series

$$f(z) = \sum_{n=1}^{\infty} \frac{1}{(z+n)^2}$$

is meromorphic on every bounded subset of $\mathbf{C}$, and find the residues at its poles. (Hint: Can you do it for a finite series?)

Solutions

1. Let $s_n(z) = 1 + 2z + 3z^2 + \cdots + nz^{n-1}$; then

$$zs_n(z) = z + 2z^2 + \cdots + (n-1)z^{n-1} + nz^n$$

and so

$$(1-z)s_n(z) = 1 + z + z^2 + \cdots + z^{n-1} - nz^n$$

$$= \frac{1 - z^n}{1 - z} - nz^n, \quad z \neq 1,$$

so that

$$s_n(z) = \frac{1 - z^n}{(1-z)^2} - \frac{nz^n}{1-z}.$$

(This can also be obtained by differentiating the (finite) geometric series.)

Hence,

$$\left| \frac{1}{(1-z)^2} - s_n(z) \right| = \left| \frac{nz^n}{1-z} + \frac{z^n}{(1-z)^2} \right| = \left| \frac{z^n(n+1) - nz^{n+1}}{(1-z)^2} \right|.$$

Now let D be any closed disc in the disc $|z| < 1$, then for $z \in D$ we have

$$\left| \frac{1}{(1-z)^2} - s_n(z) \right| = \frac{|z^n|}{|1-z|^2} \cdot |n + 1 - nz|$$

$$\leqslant \frac{\rho^n}{(1-\rho)^2} \cdot (2n+1),$$

where ρ is such that $D \subset \{z : |z| \leqslant \rho < 1\}$.

The right-hand expression has limit zero for large n; hence, given $\varepsilon > 0$, there is a number N such that $\left| \dfrac{1}{(1-z)^2} - s_n(z) \right| < \varepsilon$ if $n > N$ for all $z \in D$.

2. If $|z| \leqslant a$, then

$$|z - (a + n)| \geqslant \big||z| - |a + n|\big|$$
$$= \big||z| - (a + n)\big|$$
$$= a + n - |z|$$
$$\geqslant n.$$

Hence, $\left|\dfrac{\sin z}{z - (a + n)}\right| \leqslant \dfrac{M}{n}$, where $M = \sup\{|\sin z| : |z| \leqslant a\}$, which exists because $z \longrightarrow |\sin z|$ is continuous on the compact set $\{z : |z| \leqslant a\}$. It follows that the sequence of functions converges uniformly to the zero function on the given disc.

3. We have $|e^{nz}| = e^{nx} = (e^x)^n$, where $z = x + iy$; so that, for any $\alpha < 0$, if $x \leqslant \alpha$, then $e^x \leqslant e^\alpha < 1$. But $\displaystyle\sum_{n=1}^{\infty} e^{\alpha n}$ converges (because it is a geometric series and $e^\alpha < 1$), and so $\displaystyle\sum_{n=1}^{\infty} e^{nz}$ converges uniformly on any half-plane $\{z : \operatorname{Re} z \leqslant \alpha < 0\}$. It follows, using Theorem 3, that f is analytic on the half-plane $\{z : \operatorname{Re} z < 0\}$.

The series diverges and f is undefined if $\operatorname{Re} z \geqslant 0$ (since the terms do not then converge to zero).

4. For each z in A we must show that f is continuous at z on A. So choose any particular z in A and let $\varepsilon > 0$. Since $\{f_n\}$ converges uniformly to f on A, there must be some N such that

$$|f(w) - f_N(w)| < \frac{\varepsilon}{3} \quad \text{for all } w \text{ in } A.$$

In particular, for each h such that $z + h$ is in A, we have

$$|f(z) - f_N(z)| < \frac{\varepsilon}{3} \tag{1}$$

and

$$|f(z + h) - f_N(z + h)| < \frac{\varepsilon}{3}. \tag{2}$$

Now f_N is continuous on A, so there is some $\delta > 0$ such that for $|h| < \delta$ and $z + h \in A$, we have

$$|f_N(z) - f_N(z + h)| < \frac{\varepsilon}{3}. \tag{3}$$

It follows that

$$|f(z + h) - f(z)| = |f(z + h) - f_N(z + h) + f_N(z + h) - f_N(z) + f_N(z) - f(z)|$$
$$\leqslant |f(z + h) - f_N(z + h)| + |f_N(z + h) - f_N(z)| + |f_N(z) - f(z)|$$
$$< \frac{\varepsilon}{3} + \frac{\varepsilon}{3} + \frac{\varepsilon}{3}, \quad \text{from (1), (2) and (3)},$$
$$= \varepsilon.$$

This proves that f is continuous at z on A.

5. If $\{f_n\}$ converges uniformly to f on A then, for any $\varepsilon > 0$, there is N such that if $n > N$, then

$$|f(z) - f_n(z)| < \varepsilon \quad \text{for all } z \text{ in } A.$$

If g is continuous on the compact set A, then g is bounded on A, so that $|g(z)| < m$ (say) for all z in A. Hence, if $n > N$, we have

$$|g(z)f(z) - g(z)f_n(z)| = |g(z)| \cdot |f(z) - f_n(z)|$$
$$< m\varepsilon \quad \text{for all } z \text{ in } A,$$

which proves the required result.

6. The sequence of functions $\{z \longrightarrow z^n + 3z - 1\}$ converges uniformly to $z \longrightarrow 3z - 1$ on the disc $|z| \leqslant \rho$, where $\rho < 1$, because

$$|(z^n + 3z - 1) - (3z - 1)| = |z|^n \leqslant \rho^n.$$

Hence, given $\varepsilon > 0$, we may choose N to be any integer greater than $\dfrac{\log \varepsilon}{\log \rho}$ to ensure that $|z|^n < \varepsilon$ if $n > N$ for *all* $z \in \{z : |z| \leqslant \rho\}$.

7. The essence of this problem is that we can express f as the sum of two functions, f_1 and f_2, by writing

$$f_1(z) = \sum_{n=1}^{N} \frac{1}{(z+n)^2} \quad \text{and} \quad f_2(z) = \sum_{n=N+1}^{\infty} \frac{1}{(z+n)^2},$$

so that $f = f_1 + f_2$. We show that f_2 is analytic on the disc $D = \{z : |z| < N + \frac{1}{2}\}$, and, since f_1 has only poles in D, it follows that f is meromorphic on D. But N may be as large as we please, and therefore f is meromorphic on any bounded set.

To show that f_2 is analytic on D we notice that

$$|z + n| \geqslant |n - |z||$$
$$\geqslant n - N - \tfrac{1}{2} > 0, \quad \text{since } n \geqslant N + 1.$$

It follows that

$$\frac{1}{|z+n|^2} \leqslant \frac{1}{(n - (N + \frac{1}{2}))^2}, \quad \text{if } |z| < N + \tfrac{1}{2},$$

but

$$\sum_{n=N+1}^{\infty} \frac{1}{(n - N - \frac{1}{2})^2} = \sum_{n=1}^{\infty} \frac{1}{(n - \frac{1}{2})^2}$$

is convergent, and the required result follows from the Weierstrass M-test.

11.3 DEFINITION OF ANALYTIC FUNCTIONS BY MEANS OF INTEGRALS

We return now to our discussion of "differentiation under the integral sign". From *Unit 5*, *Cauchy's Theorem I*, we have Cauchy's Formula

$$f(z) = \frac{1}{2\pi i} \int_C \frac{f(\zeta)}{\zeta - z} \, d\zeta,$$

(where C is a circle, and f is analytic on some region R containing C and its inside) and Cauchy's Formula for the first derivative

$$f'(z) = \frac{1}{2\pi i} \int_C \frac{f(\zeta)}{(\zeta - z)^2} \, d\zeta.$$

We motivated the proof of Cauchy's formula for the first derivative (in Section 5.5 of *Unit 5*) by considering the function H of two complex variables, defined by

$$H(\zeta, z) = \frac{f(\zeta)}{\zeta - z}.$$

Then

$$f(z) = \frac{1}{2\pi i} \int_C H(\zeta, z) \, d\zeta,$$

and, as we said in *Unit 5*, it seems plausible that

$$f'(z) = \frac{1}{2\pi i} \int_C \frac{\partial H}{\partial z}(\zeta, z) \, d\zeta,$$

and indeed we proved that this result *is* true. In this case

$$\frac{\partial H}{\partial z}(\zeta, z) = \frac{f(\zeta)}{(\zeta - z)^2}.$$

This together with Example 6 of Section 11.1 gives us good reason to believe that we might be able to prove a general result: this is indeed the case. But first a preliminary problem.

Preliminary Problem

If $H:(w, z) \longrightarrow z^2[1 + \exp(zw)]$, specify

(i) $w \longrightarrow H(w, z_0)$, where z_0 is fixed;

(ii) $z \longrightarrow H(w_0, z)$, where w_0 is fixed;

(iii) $\dfrac{\partial H}{\partial z}(w, z).$

Solution

(i) $w \longrightarrow z_0^2[1 + \exp(z_0 w)]$;

(ii) $z \longrightarrow z^2[1 + \exp(zw_0)]$;

(iii) $z^2 \cdot w \exp(zw) + 2z \cdot [1 + \exp(zw)]$.

(Note that if, for each $z_0 \in B$, $w \longrightarrow z_0^2[1 + \exp(z_0 w)]$ has domain A and, for each $w_0 \in A$, $z \longrightarrow z^2[1 + \exp(zw_0)]$ has domain B then the domain of H is the Cartesian product $A \times B$.)

Theorem 5

Let R be a region and Γ a contour (not necessarily in R), and let $H : \Gamma \times R \longrightarrow \mathbb{C}$ be bounded on $\Gamma \times R$.

If (i) for each fixed $z_0 \in R$ the function

$$w \longrightarrow H(w, z_0) \text{ is continuous on } \Gamma,$$

and (ii) for each fixed $w_0 \in \Gamma$ the function

$$z \longrightarrow H(w_0, z) \text{ is analytic on } R,$$

then the function ϕ defined by

$$\phi(z) = \int_\Gamma w \longrightarrow H(w, z) = \int_\Gamma H(w, z)\, dw$$

is analytic on R, and $\phi'(z) = \int_\Gamma \dfrac{\partial H}{\partial z}(w, z)\, dw.$

Proof

The proof is similar to that for Cauchy's Formula for the first derivative (Theorem 5 of *Unit 5*).

Choose an arbitrary point z of R; then, using condition (ii) and Cauchy's Formula, we have, for each point $w \in \Gamma$,

$$H(w, z) = \frac{1}{2\pi i} \int_C \frac{H(w, \zeta)}{(\zeta - z)}\, d\zeta,$$

where C is a circle with centre z and radius ρ sufficiently small for the closed disc $\{\zeta : |\zeta - z| \leqslant \rho\}$ to lie in R (Fig. 5).

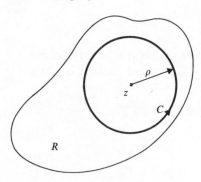

Fig. 5

Then, since we wish to show that ϕ is differentiable, we consider

$$\frac{\phi(z + h) - \phi(z)}{h} = \int_\Gamma \frac{H(w, z + h) - H(w, z)}{h}\, dw,$$

$$= \int_\Gamma \left[\frac{1}{2\pi i} \int_C \frac{H(w, \zeta)}{\zeta - z - h}\, d\zeta - \frac{1}{2\pi i} \int_C \frac{H(w, \zeta)}{\zeta - z}\, d\zeta \right] \frac{1}{h}\, dw,$$

provided that $|h| < \rho$.

Now, repeating one of the steps in the proof of Cauchy's Formula for the first derivative, we notice that

$$\frac{1}{h}\left(\frac{1}{\zeta - z - h} - \frac{1}{\zeta - z} \right) = \frac{1}{(\zeta - z - h)(\zeta - z)}$$

$$= \frac{1}{(\zeta - z)^2} + \frac{h}{(\zeta - z)^2(\zeta - z - h)},$$

so that

$$\frac{\phi(z+h) - \phi(z)}{h} = \int_{\Gamma} \left[\frac{1}{2\pi i} \int_C \frac{H(w, \zeta)}{(\zeta - z)^2} \, d\zeta \right] dw$$

$$+ \int_{\Gamma} \left[\frac{1}{2\pi i} \int_C \frac{hH(w, \zeta)}{(\zeta - z)^2(\zeta - z - h)} \, d\zeta \right] dw. \qquad (1)$$

We now intend to show that the first of the integrals on the right gives us the value that we want for $\phi'(z)$, and the other is small in modulus when $|h|$ is small.

We have

$$\frac{1}{2\pi i} \int_C \frac{H(w, \zeta)}{(\zeta - z)^2} \, d\zeta = \frac{\partial H}{\partial z}(w, z)$$

from Cauchy's Formula for the first derivative, and

$$\left| \frac{1}{2\pi i} \int_C \frac{hH(w, \zeta)}{(\zeta - z)^2(\zeta - z - h)} \, d\zeta \right| \leqslant \frac{1}{2\pi} \cdot \frac{|h|K \cdot 2\pi\rho}{\rho^2(\rho - |h|)},$$

from the Estimation Theorem, and the fact that H is bounded on $\Gamma \times R$ (by K, say).

Then, using the Estimation Theorem again, we have

$$\left| \int_{\Gamma} \left[\frac{1}{2\pi i} \int_C \frac{hH(w, \zeta)}{(\zeta - z)^2(\zeta - z - h)} \, d\zeta \right] dw \right| \leqslant \frac{1}{2\pi} \cdot \frac{|h|K \cdot 2\pi\rho}{\rho^2(\rho - |h|)} \cdot L$$

$$= \frac{|h|KL}{\rho(\rho - |h|)},$$

where L is the length of Γ. Notice that the final expression is small when $|h|$ is small. It follows from (1) that

$$\phi'(z) = \lim_{h \to 0} \frac{\phi(z+h) - \phi(z)}{h}$$

$$= \int_{\Gamma} \frac{\partial H}{\partial z}(w, z) \, dw,$$

as required. ∎

Example 1

We now return to Example 6, of Section 11.1, where $H(w, z) = e^{zw}f(w)$ and Γ is the line segment $[0, 1]$. We have to verify only that the conditions of Theorem 5 hold. If we choose the region R of that theorem to be any open disc $\{z : |z| < r\}$, then, for $(w, z) \in \Gamma \times R$

$$|H(w, z)| \leqslant e^{|zw|}|f(w)|$$

$$\leqslant e^r M, \quad \text{where } M = \sup\{|f(z)| : z \in [0, 1]\},$$

so that H is bounded, as required.

Then (i) for all fixed $z_0 \in R$, the function $w \longrightarrow \exp(z_0 w)f(w)$ is continuous on $[0, 1]$, and (ii) for all fixed $w_0 \in [0, 1]$, the function $z \longrightarrow \exp(z w_0)f(w_0)$ is entire.

Thus the conditions of the theorem hold and we deduce that the function ϕ defined by $\phi(z) = \int_0^1 e^{zw}f(w) \, dw$ is analytic on every open disc $|z| < r$, and, moreover, that

$$\phi'(z) = \int_0^1 we^{zw}f(w) \, dw.$$

Example 2

It is possible to extend Theorem 5 to "infinite contours" and in this example we show how this can be done for "straight line contours". The general case is very similar.

Consider the integral $\int_0^k H(w, z)\, dw$, $k \in \mathbf{R}$, where H satisfies the conditions of Theorem 5 for some region R and where Γ of that theorem is taken to be any line segment $[0, k]$.

Suppose now that $\lim_{k \to \infty} \int_0^k H(w, z)\, dw$ exists, and, moreover, that the convergence is uniform on any closed disc in R, in the sense that: given $\varepsilon > 0$, there exists a number K (independent of z in the disc) and a function ϕ such that

$$\left| \phi(z) - \int_0^k H(w, z)\, dw \right| < \varepsilon \quad \text{if } k > K.$$

We then say that the integral $\int_0^\infty H(w, z)\, dw$ *converges uniformly* to ϕ on every closed disc in R.

We now form a sequence of functions $\{\phi_n\}$ by putting $\phi_n(z) = \int_0^n H(w, z)\, dw$. This sequence converges uniformly to ϕ on any closed disc in R, and hence, from Theorem 3 and Theorem 5, we deduce that ϕ is analytic on R and, moreover, $\lim_{n \to \infty} \phi'_n = \phi'$.

We have taken the limit through *integer* values but we must now allow k to range through any values in $\mathbf{R}$. In fact,

$$\lim_{k \to \infty} \int_{[k]} H(w, z)\, dw = 0,$$

where $[k]$ is the largest integer $\leq k$, as a consequence of the convergence of the integral. Finally we deduce that, under these conditions,

$$\frac{d}{dz} \int_0^\infty H(w, z)\, dw = \int_0^\infty \frac{\partial H}{\partial z}(w, z)\, dw.$$

Summary

In this section we have proved a theorem which enables us to determine whether certain functions defined in terms of integrals are analytic.

Self-Assessment Question

State conditions which ensure that if

$$\phi(z) = \int_\Gamma H(w, z)\, dw,$$

then

$$\phi'(z) = \int_\Gamma \frac{\partial H}{\partial z}(w, z)\, dw,$$

where Γ is a contour to be specified.

Solution

See Theorem 5.

11.4 PROBLEMS

1. Show that if f is continuous on a contour Γ then the function ϕ defined by

$$\phi(z) = \int_\Gamma \frac{f(w)}{w - z} \, dw$$

    is analytic on any region which lies in a compact set not containing points of Γ.

2. If f is a function continuous on the interval $[-1, 1]$, show that the function ϕ defined by

$$\phi(z) = \int_{-1}^{1} f(t) \sin(zt) \, dt$$

    is entire.

3. If ϕ is a function analytic on a region R and Γ is a closed contour in R such that $|\phi(z)| < 1$ if $z \in \Gamma$, expand $\dfrac{1}{1 + \phi(z)}$ in powers of $\phi(z)$ and prove that

$$\int_\Gamma \frac{\phi'}{1 + \phi} = 0.$$

    (This result will be used in Section 11.8 in the proof of Rouché's Theorem.)

Solutions

1. Let R be a region lying in a compact set Δ which has no points in common with Γ. Let d be the least distance from any point of Γ to any point of Δ and let $M = \sup\{|f(w)| : w \in \Gamma\}$, then

$$\left| \frac{f(w)}{w - z} \right| \leqslant \frac{M}{d} \quad \text{if } (w, z) \in \Gamma \times R.$$

    Also, for each fixed $z_0 \in R$, the function

$$w \longrightarrow \frac{f(w)}{w - z_0} \quad \text{is continuous on } \Gamma,$$

    and, for each fixed $w_0 \in \Gamma$, the function

$$z \longrightarrow \frac{f(w_0)}{w_0 - z} \quad \text{is analytic on } R.$$

    We may now apply Theorem 5 to obtain the required result.

2. Let H be defined by $H(w, z) = f(w) \sin(zw)$, let $\Gamma = [-1, 1]$ and let R be any bounded region. Also let $M = \sup\{|f(w)| : w \in [-1, 1]\}$, which exists because $|f|$ is continuous on $[-1, 1]$.

    Since R is bounded it lies in some disc $\{z : |z| \leqslant a\}$, and therefore

$$\begin{aligned}
|\sin(zw)| &= \left| \frac{e^{izw} - e^{-izw}}{2i} \right| \\
&\leqslant \tfrac{1}{2}(|e^{izw}| + |e^{-izw}|) \\
&\leqslant e^{|zw|} \\
&\leqslant e^a, \quad \text{since } |z| \leqslant a \text{ and } w \in [-1, 1].
\end{aligned}$$

    It follows that $|H(z, w)| \leqslant Me^a$, so that H is bounded on $\Gamma \times R$. For each fixed $z_0 \in R$, the function $w \longrightarrow f(w) \sin(z_0 w)$ is continuous on Γ, and, for each fixed $w_0 \in \Gamma$, the function $z \longrightarrow f(w_0) \sin(zw_0)$ is analytic on R (since sin is entire).

    The conditions of Theorem 5 are satisfied, and it follows that

$$\phi : z \longrightarrow \int_{-1}^{1} f(t) \sin(zt) \, dt$$

    is analytic on every bounded region R, and hence ϕ is entire.

3. The function ϕ is continuous on the compact set Γ and therefore $|\phi|$ attains its upper bound k (say) at some point z_0 of Γ. Clearly, $k < 1$ (for otherwise $|\phi(z_0)| \geq 1$), so that

$$|\phi(z)| \leq k < 1, \quad \text{if } z \in \Gamma.$$

For each $z \in \Gamma$, we have

$$\frac{1}{1 + \phi(z)} = \sum_{n=0}^{\infty} [-\phi(z)]^n,$$

since the right-hand side is just the sum of a geometric series. But this series is uniformly convergent on Γ, by the Weierstrass M-test, and we may therefore use Theorem 1 to obtain

$$\int_{\Gamma} \frac{\phi'(z)}{1 + \phi(z)} dz = \sum_{n=0}^{\infty} \int_{\Gamma} [-\phi(z)]^n \phi'(z) \, dz.$$

Notice now that

$$[-\phi(z)]^n [-\phi'(z)] = \frac{d}{dz} \frac{[-\phi(z)]^{n+1}}{n+1},$$

so that each of the integrals on the right is zero from the Fundamental Theorem (*Unit 4*), Γ being a closed contour.

(Note that the result may be established more simply. Let $w = \phi(z)$; then

$$\int_{\Gamma} \frac{\phi'(z)}{1 + \phi(z)} dz = \int_{\phi(\Gamma)} \frac{1}{1 + w} dw$$

$$= 0$$

since $\phi(\Gamma)$ is a closed contour in the region $\{z : |z| < 1\}$ and $w \longrightarrow \dfrac{1}{1 + w}$ is analytic on this region.)

11.5 ANALYTIC CONTINUATION

The treatment in this section must be somewhat intuitive, since we have neither the time nor the necessary topological background to develop the theory of analytic continuation in full detail. However, you will see that it is possible to tie together certain aspects of the course provided that you are ready to accept this less formal approach, and perhaps regard it as an introduction to a more advanced study later in your career. Some ideas from this section will also be useful for certain parts of the later units.

You will have noticed that so far in this course we have not paid a great deal of attention to the domains of the functions under consideration. For general discussions involving a function f on a region R there has always been an implicit assumption that R is a subset of the domain of f. In particular cases where the function f is defined by a formula, we have taken the domain to be the largest set on which the formula makes sense. For example:

(i)   if $\phi_1(z) = \sum_{n=0}^{\infty} z^n$, then the domain of ϕ_1 is the set $\{z : |z| < 1\}$;

(ii) if $\phi_2(z) = \dfrac{1}{1-z}$, then the domain of ϕ_2 is the set $\mathbf{C} - \{1\}$;

(iii) if $\phi_3(z) = \lim_{r \to \infty} \int_0^r e^{-t(1-z)} \, dt$, then the domain of ϕ_3 is the set $\{z : \operatorname{Re} z < 1\}$,

   for it is only on this set that the limit exists.

However, it is not difficult to see that $\phi_1(z) = \dfrac{1}{1-z}$ if $|z| < 1$, and $\phi_3(z) = \dfrac{1}{1-z}$ if $\operatorname{Re} z < 1$, so that, in a sense, ϕ_1, ϕ_2 and ϕ_3 are the "same" function. In this section we shall attempt to give some meaning to this statement.

You may feel that the restriction of the domains of ϕ_1 and ϕ_3 is a little artificial, that their "natural domain" is the set $\mathbf{C} - \{1\}$, and it is only technical difficulties with sums and integrals which cause the trouble. You would be right to think in this way, for we are going to show that every function f analytic on a region R has a "natural domain" which may well contain R as a *proper* subset. This may lead you to suspect that we intend to look for the "largest" set in $\mathbf{C}$ which will serve as the domain of f. In a sense, this is right, for we set out to increase the size of the domain as much as possible, but we shall encounter the difficulty that some functions do not have a largest domain—at least, not in a simple sense. But let us say no more at this stage, and wait for the difficulty to arise.

We should like some general method of determining the "natural domain" of a function. If all functions could be specified by a formula as easily as ϕ_1 and ϕ_3 can, then our problem would soon be resolved. Unfortunately, that is not the case, and we might be presented with functions such as

$$\zeta(z) = \sum_{n=1}^{\infty} n^{-z} \qquad \text{(the Riemann zeta function)},$$

$$\Gamma(z) = \int_0^{\infty} e^{-t} t^{z-1} \, dt \qquad \text{(the gamma function)},$$

and

$$F(z) = \sum_{n=0}^{\infty} z^{2^n}.$$

It is by no means easy to guess a "natural domain" for such functions, or even to see how such a natural domain could be defined. We are about to discuss the process of analytic continuation which enables us to define such a "natural domain", but the determination of the "natural domain" of a particular function may well be extremely difficult and certainly beyond the scope of this course, except in a few relatively simple cases.

Direct Analytic Continuation

As the first step towards the formulation of a suitable definition of the "natural domain" of a function f which is known to be analytic on a region R, we shall attempt to find a natural extension of f to a region larger than R. For this purpose we shall require the following concept.

Definition

> If a function is analytic on a region R, and R is the domain of f, then the pair (f, R) is called a **function element**.

The extension of f which we are trying to construct will appear as a collection of function elements which are related to each other in a prescribed fashion.

Definition

> Two function elements (f_1, R_1) and (f_2, R_2) are said to be **direct analytic continuations of each other** if $f_1(z) = f_2(z)$ for all z in some non-empty open subset of $R_1 \cap R_2$. We also say that (f_2, R_2) is a **direct analytic continuation of** (f_1, R_1) **from** R_1 **onto** R_2.

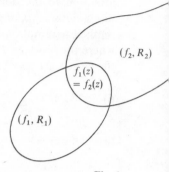

Fig. 6

For example, if $\phi_1(z) = \sum_{n=0}^{\infty} z^n$, $R_1 = \{z : |z| < 1\}$, and $\phi_2(z) = \dfrac{1}{1-z}$, $R_2 = \mathbb{C} - \{1\}$, then (ϕ_2, R_2) is a direct analytic continuation of (ϕ_1, R_1) from R_1 onto R_2. In this case R_1 is a proper subset of R_2 (Fig. 7), and we might regard ϕ_2 as the "natural" extension of ϕ_1 from R_1 onto R_2. But it is sensible to talk of *the* extension of ϕ_1 from R_1 onto R_2 only if this extension is unique.

The process of direct analytic continuation is indeed unique, and, using the Uniqueness Theorem of *Unit 6*, we can easily verify that this is so. For convenience we repeat the Uniqueness Theorem:

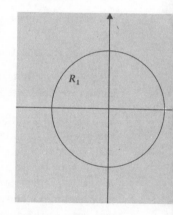

Fig. 7

The Uniqueness Theorem   (Theorem 16 of Unit 6)

Let f and g be analytic on a region R, let S be a subset of R with a cluster point in R, and let $f = g$ on S. Then $f = g$ on R.

If (f_2, R_2) and (g, R_2) are both direct analytic continuations of (f_1, R_1), and $R_1 \cap R_2$ is a region, then we can show that $g = f_2$ (Fig. 8). Applying the Uniqueness Theorem to f_1 and f_2 on the region $R_1 \cap R_2$ we deduce that $f_1(z) = f_2(z)$ for $z \in R_1 \cap R_2$. Similarly, $f_1(z) = g(z)$ for $z \in R_1 \cap R_2$, and therefore $f_2 = g$ on $R_1 \cap R_2$. Using the Uniqueness Theorem again, we have $f_2 = g$ on R_2. In other words, if $R_1 \cap R_2$ is a region, then there is only one direct analytic continuation of f_1 from R_1 onto R_2, and we have proved the following theorem.

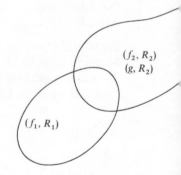

Fig. 8

Theorem 6   (The Uniqueness of Direct Analytic Continuation)

If the regions R_1 and R_2 are such that $R_1 \cap R_2$ is a region and if the two function elements (f_2, R_2) and (g, R_2) are both direct analytic continuations of (f_1, R_1), then $f_2 = g$.

If (f_2, R_2) is a direct analytic continuation of (f_1, R_1) and $R_1 \cap R_2$ is a region then we may define a function element $(f, R_1 \cup R_2)$ by setting

$$f(z) = \begin{cases} f_1(z), & z \in R_1 \\ f_2(z), & z \in R_2. \end{cases}$$

If we can find a direct analytic continuation (f_2, R_2) of (f_1, R_1) such that R_1 is a *proper* subset of $R_1 \cup R_2$, then we can regard the process of extending (f_1, R_1) to $(f, R_1 \cup R_2)$ as successful. Having extended the domain from R_1 to $R_1 \cup R_2$, there appears to be no reason why we should not repeat the process and so push the domain out to its furthest extent. This is certainly our intention, but, as we shall see later, there are some difficulties to overcome. Before we discuss the further extension of the domain we shall point out a very important property of direct analytic continuation, but first we give an illustration of the idea.

Example 1

Suppose that (f, R_1) is a *solution* of the differential equation

$$\phi''(z) + (z^2 - 1)\phi'(z) + \phi(z) = \sin z,$$

by which we mean

$$f''(z) + (z^2 - 1)f'(z) + f(z) = \sin z \quad \text{for } z \in R_1,$$

and let (g, R_2) be a direct analytic continuation of (f, R_1). Is (g, R_2) also a solution of the differential equation?

The answer is "yes" and it is not difficult to establish that this is the case. For, if (g, R_2) is a direct analytic continuation of (f, R_1), then (g', R_2) and (g'', R_2) are the corresponding direct analytic continuations of (f', R_1) and (f'', R_1), respectively. (We ask you to prove this in the next set of Self-Assessment Questions.)

If we now define functions F_1 and F_2 by

$$F_1(z) = f''(z) + (z^2 - 1)f'(z) + f(z) - \sin z, \quad z \in R_1,$$

and

$$F_2(z) = g''(z) + (z^2 - 1)g'(z) + g(z) - \sin z, \quad z \in R_2,$$

then (F_2, R_2) is a direct analytic continuation of (F_1, R_1). But $F_1 = 0$ on R_1 and therefore $F_2 = 0$ on R_2 (by the Uniqueness Theorem), which completes the argument.

A similar discussion will establish a general result, which, although lengthy to state, follows immediately from the Uniqueness Theorem.

Theorem 7 (The Permanence of Functional Relationships)

Let $G: \mathbf{C}^n \longrightarrow \mathbf{C}$ and let $(f_1, R), (f_2, R), \ldots, (f_n, R)$, be a set of function elements such that

$$G(f_1(z), f_2(z), \ldots, f_n(z)) = 0 \text{ for all } z \text{ on a subset } S \text{ of } R \text{ with a cluster point in } R.$$

If $z \longrightarrow G(f_1(z), f_2(z), \ldots, f_n(z))$ is analytic on R then it must be the zero function on R.

The following example shows how Theorem 7 may be used to extend an appropriate real analysis result (namely: $\sin 2x = 2 \sin x \cos x$) to complex analysis.

Example 2

In the notation of Theorem 7, let $G: \mathbf{C}^3 \longrightarrow \mathbf{C}$ be given by

$$G(z_1, z_2, z_3) = z_1 - 2z_2 z_3;$$

let

$$f_1: z \longrightarrow \sin 2z, \quad z \in \mathbf{C},$$

$$f_2: z \longrightarrow \sin z, \quad z \in \mathbf{C},$$

$$f_3: z \longrightarrow \cos z, \quad z \in \mathbf{C},$$

(So R of Theorem 7 is $\mathbf{C}$.)

Then, since $\sin 2x = 2 \sin x \cos x, x \in \mathbf{R}$, we have

$$G(f_1(z), f_2(z), f_3(z)) = \sin 2z - 2 \sin z \cos z$$

$$= 0 \text{ for all } z \text{ in } \mathbf{R}.$$

83

The set $\mathbf{R}$ is a subset of $\mathbf{C}$ and has a cluster point in $\mathbf{C}$. (So S of Theorem 7 is $\mathbf{R}$.)

Also

$$z \longrightarrow G(f_1(z), f_2(z), f_3(z)) = \sin 2z - 2 \sin z \cos z$$

is analytic on $\mathbf{C}$ and so, by Theorem 7, is the zero function on $\mathbf{C}$, that is

$$\sin 2z = 2 \sin z \cos z, \quad z \in \mathbf{C}.$$

Self-Assessment Questions

1. If (f_1, R_1) is a direct analytic continuation of (f_2, R_2), and $f_1 = 0$ on R_1, show that $f_2 = 0$ on R_2.

2. If (f_1, R_1) is a direct analytic continuation of (f_2, R_2), show that (f_1', R_1) is the corresponding direct analytic continuation of (f_2', R_2).

3. If $R = \{z : \operatorname{Re} z > 0\}$, (f, R) is a function element, and $f(z + 1) = f(z)$ for all $z \in R$, what can you say about f? Is there a direct analytic continuation of f onto a region larger than R?

Solutions

1. The function f defined by

    $$f(z) = \begin{cases} f_1(z), & z \in R_1 \\ f_2(z), & z \in R_2, \end{cases}$$

    is analytic on $R_1 \cup R_2$ and $f = 0$ on the set $R_1 \cap R_2 \neq \varnothing$. It follows from the Uniqueness Theorem of *Unit 6* that $f = 0$ on $R_1 \cup R_2$, and therefore $f_2 = 0$ on R_2.

2. We know that $f_1 = f_2$ on $R_1 \cap R_2$, and therefore $f_1' = f_2'$ on $R_1 \cap R_2$, so that (f_1', R_1) is a direct analytic continuation of (f_2', R_2).

3. The function f is periodic with period 1. Let $R_1 = \{z : \operatorname{Re} z > -1\}$ and consider the function $f_1 : z \longrightarrow f(z + 1), z \in R_1$. Notice that $f_1(z) = f(z + 1) = f(z)$ if $z \in R$ and that $R \cap R_1 = R$ is nonempty, so that (f_1, R_1) is a direct analytic continuation of (f, R). Thus, the answer to our question is "yes".

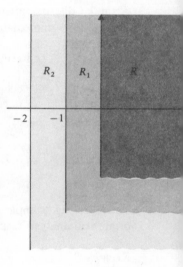

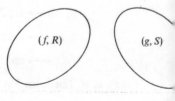

Analytic Multi-Functions

Notice that for the function element (f_1, R_1) in Self-Assessment Question 3 we can use the permanence of functional relationships to show that $f_1(z + 1) = f_1(z)$. It follows that the same argument can be used to find a direct analytic continuation (f_2, R_2) of (f_1, R_1): we need only choose

$$f_2(z) = f_1(z + 1), \quad z \in R_2 = \{z : \operatorname{Re} z > -2\}.$$

Fig. 9

This process could be continued indefinitely, so that there is an entire function ϕ such that $(\phi, \mathbf{C})$ is a direct analytic continuation of (f, R) and, moreover, $\phi(z + 1) = \phi(z), z \in \mathbf{C}$. In this case it would be very reasonable to regard $(\phi, \mathbf{C})$ as the extension of (f, R) obtained by making the domain as large as possible. In a sense, f and ϕ are the "same" function, the only difference being that the domain of ϕ is larger than that of f.

In general, if (f_2, R_2) is a direct analytic continuation of (f_1, R_1) then it is very reasonable to regard f_1 and f_2 as the "same" function. But suppose that we are given two function elements (f, R) and (g, S) such that $S \cap R$ is empty (Fig. 10). What then?

Fig. 10

For example, let

$$f(z) = \sum_{n=0}^{\infty} z^n, \quad R = \{z : |z| < 1\}$$

and

$$g(z) = - \sum_{n=0}^{\infty} (2 - z)^n, \quad S = \{z : |z - 2| < 1\};$$

then $S \cap R = \varnothing$ (Fig. 11).

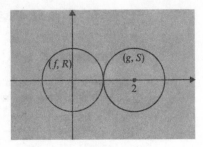

Fig. 11

If you notice that $f(z) = \dfrac{1}{1 - z}$, $z \in R$, and $g(z) = \dfrac{1}{1 - z}$, $z \in S$, then presumably there is no difficulty, and you would not hesitate to say that f and g are the "same" function. We must say what we mean by the word "same"; in order to do this we introduce the following definitions.

Definition

> The finite sequence of function elements
>
> $$\{(f_1, R_1), (f_2, R_2), \ldots, (f_n, R_n)\}$$
>
> is said to form a **chain** if (f_{k+1}, R_{k+1}) is the direct analytic continuation of (f_k, R_k) for $k = 1, 2, \ldots, n - 1$. The elements of the chain are said to be **analytic continuations of each other**, and the chain is said to **join** the function elements (f_1, R_1) and (f_n, R_n). If $R_1 = R_n$, the chain is said to be **complete**, and if $R_k \subset R$ for $k = 1, 2, \ldots, n$, then we say that the chain is **contained in R**.

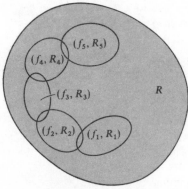
A chain contained in R

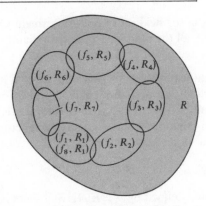
A complete chain contained in R

Fig. 12

We can use the concept of a chain of function elements to construct an equivalence relation (see *Unit M100 19, Relations*) on the set of all function elements, for we may write

$$(f, R) \text{ is related to } (g, S)$$

if there is a chain joining (f, R) and (g, S). (In the next set of Self-Assessment Questions we ask you to prove that this *is* an equivalence relation.) The corresponding equivalence classes are called **analytic multi-functions** and the elements of an equivalence class are sometimes called the **branches** of the analytic multi-function. (The words "branch" and "function element" are used interchangeably. In particular, this use of "branch" is consistent with our use of the term in Section 3.6 of *Unit 3, Differentiation*.) The process of constructing chains from function elements to a given function element (f, R) is called **analytic continuation** and it is the process by which, in theory, we can construct the analytic multi-function to which (f, R) belongs.

An analytic multi-function is not a function in our usual sense, but a whole collection of function elements—an equivalence class. For example, if $\phi_1(z) = \sum_{n=0}^{\infty} z^n$, $R_1 = \{z : |z| < 1\}$, then the corresponding multi-function is the set of *all* function elements which can be joined to (ϕ_1, R_1) by a chain. Here are three more elements of this multi-function.

$$\phi_2 : z \longrightarrow \frac{1}{1-z}, \quad R_2 = \mathbf{C} - \{1\};$$

$$\phi_3 : z \longrightarrow \int_0^{\infty} e^{-t(1-z)}\, dt, \quad R_3 = \{z : \operatorname{Re} z < 1\};$$

$$\phi_4 : z \longrightarrow -\sum_{n=0}^{\infty} (2-z)^n, \quad R_4 = \{z : |z - 2| < 1\}.$$

Notice that $R_1 \cap R_4 = \varnothing$ (Fig. 13), but there is a chain $\{(\phi_1, R_1), (\phi_2, R_2), (\phi_4, R_4)\}$ joining (ϕ_1, R_1) and (ϕ_4, R_4), so that they belong to the same analytic multi-function.

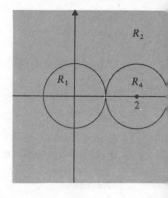

Fig. 13

From the permanence of functional relationships we have, for any function element (ϕ, R) of this analytic multi-function,

$$\phi(z) = \frac{1}{1-z}, \quad z \in R \quad \text{(since } \phi(z)(1-z) = 1\text{)}.$$

Of course, for this analytic multi-function there is a function element with largest possible domain, namely

$$z \longrightarrow \frac{1}{1-z}, \quad z \in \mathbf{C} - \{1\}.$$

(We ask you to prove this in the next set of Self-Assessment Questions.) Once we have found this element, all the others become redundant, and we can stop messing about with chains and so on.

The above discussion might lead you to suspect that, given a function element of an analytic multi-function, we need find only the function element with largest domain and then all our worries are over. Unfortunately, it is not that easy. Some analytic multi-functions do not contain a function element with largest domain, as we shall soon see.

Self-Assessment Questions

4. Show that the relation ρ defined by "$(f, R)\rho(g, S)$ if there is a chain joining (f, R) and (g, S)" is an equivalence relation.

5. Show that there is no direct analytic continuation $(f, \mathbf{C})$ of

$$\left(z \longrightarrow \frac{1}{1-z}, \mathbf{C} - \{1\} \right).$$

Solutions

4. The three requirements for an equivalence relation are satisfied:

    (a) *Reflexive* $(f, R)\rho(f, R)$ since (f, R) is a direct analytic continuation of itself from the definition;

    (b) *Symmetric* $(f, R)\rho(g, S) \Rightarrow (g, S)\rho(f, R)$, from the definition;

    (c) *Transitive* $(f, R)\rho(g, S)$ and $(g, S)\rho(h, T) \Rightarrow (f, R)\rho(h, T)$ since the "union" of two chains is also a chain.

5. Suppose that $(f, \mathbf{C})$ is a direct analytic continuation of $\Big(z \longrightarrow \frac{1}{1-z}$, $\mathbf{C} - \{1\}\Big)$; then f is continuous on the disc $|z - 1| \leqslant 1$ and therefore

$|f| < K$ on this disc, for some constant K. Hence, $|f| < K$ on $D = \{z : 0 < |z - 1| \leq 1\}$, but $f(z) = \dfrac{1}{1 - z}$ if $z \in \mathbf{C} - \{1\}$ and f is therefore unbounded on the punctured disc D, and so we have a contradiction.

The Logarithm

You should glance through Section 3.6 of *Unit 3* and the part of Section 9.3 of *Unit 9* which follows the example there, before proceeding with this sub-section.

The aim of this section is to show you that there are analytic multi-functions, such as the logarithm, which do not have a function element with largest domain.

Preliminary Problems

1. What is the domain of Log?

2. Calculate $\text{Log}(1 - i)$.

3. In the notation of Section 3.6 of *Unit 3*, how is $L_R(z)$ related to Log z, where R is the region $\{z : 0 < \text{Im } z < 2\pi\}$?

4. Define a branch of the logarithm on the simply-connected region $\{z : -\pi < \text{Arg } z < \pi\}$.

Solutions

1. Log has domain $\{z : z \neq 0$ or a negative real number$\}$, which we could equally write as $\{z : -\pi < \text{Arg } z < \pi\}$.

2. We have $\text{Log } z = \log |z| + i \text{ Arg } z$, $-\pi < \text{Arg } z < \pi$, so that

$$\text{Log}(1 - i) = \log |1 - i| + i\left(-\frac{\pi}{4}\right) = \frac{1}{2}\log 2 - i\frac{\pi}{4}.$$

3.

$$L_R(z) = \begin{cases} \text{Log } z, & \text{if } 0 < \text{Arg } z < \pi \\ \text{Log } z + 2\pi i, & \text{if } -\pi < \text{Arg } z < 0; \end{cases}$$

so that L_R agrees with Log on the upper half-plane. (See the discussion following Fig. 20 of Section 3.6 of *Unit 3*.)

4. From *Unit 9*, Section 9.3, we know that a branch of the logarithm can be defined on any simply-connected region not containing zero. For example, we could choose the principal branch

$$\text{Log } z = \log |z| + i \text{ Arg } z, \quad -\pi < \text{Arg } z < \pi.$$

There are many other branches which we could define, for example

$$z \longrightarrow \log |z| + 2k\pi i + i \text{ Arg } z, \quad -\pi < \text{Arg } z < \pi,$$

is a branch of the logarithm for any fixed integer k.

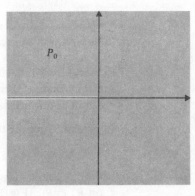

If, for convenience, we put $P_0 = \{z : -\pi < \text{Arg } z < \pi\}$ (Fig. 14), then the principal branch of the logarithm, (Log, P_0), is a function element. The analytic multi-function generated from this function element by the process of analytic continuation is called **the logarithm**. (In the notation of the Foundation Course the logarithm is the equivalence class, $[(\text{Log}, P_0)]$.)

Fig. 14

Now we ask the question: "Does the logarithm contain a function element with domain P_{max}, say, such that the domain of every other function element lies in P_{max}?"

At first sight (Log, P_0) looks like a good candidate for the required function element, since the closure of P_0 is $\mathbf{C}$. But consider the branch of the logarithm defined by

$$L(z) = \log|z| + i\theta(z)$$

where $\theta(z)$ is the argument of z such that $0 < \theta(z) < 2\pi$. (This is the function L_R of Preliminary Problem 3 with the suffix omitted for convenience.) If $P_1 = \{z : 0 < \theta(z) < 2\pi\}$, then P_1 is certainly not a subset of P_0, so we must reject (Log, P_0) as the function element of the logarithm with largest domain (Fig. 15).

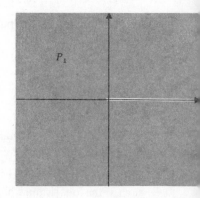

Fig. 15

Suppose now that there is a function element of the logarithm (f, P_{max}) such that $P_0 \subset P_{max}$ and $P_1 \subset P_{max}$; this will lead us to a contradiction. Since $P_0 \cup P_1 = \mathbf{C} - \{0\}$, it follows that f is analytic on $\mathbf{C} - \{0\}$. But $f'(z) = 1/z$ for $z \in \mathbf{C} - \{0\}$ from the permanence of functional relationships, so that, if Γ is a simple-closed contour enclosing the origin, then

$$\int_\Gamma f'(z)\, dz = \int_\Gamma \frac{dz}{z}.$$

But $\int_\Gamma \dfrac{dz}{z} = 2\pi i$ and $\int_\Gamma f'(z)\, dz = 0$ by the Fundamental Theorem for contour integrals (Theorem 8 of *Unit 4*). This is a contradiction.

We conclude that there is no function element of the logarithm with domain P_{max} such that the domain of every other function element of the logarithm lies in P_{max}.

It seems that our proposal for the definition of the "natural" domain of a given function element has come to grief for the logarithm. However, all is not lost: a further investigation of the properties of the logarithm will lead us in the right direction.

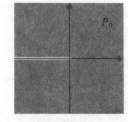

Fig. 16

The following set of function elements (in the above notation) is a complete chain (Fig. 16)

$$\{(\text{Log}, P_0), (L, P_1), (z \longrightarrow \text{Log}\, z + 2\pi i, P_0)\},$$

for (L, P_1) is a direct analytic continuation of (Log, P_0), since, as we pointed out in the solution to Preliminary Problem 3, $L(z) = \text{Log}\, z$ on the upper-half plane. (We ask you to show that $(z \longrightarrow \text{Log}\, z + 2\pi i, P_0)$ is a direct analytic continuation of (L, P_1) in the following set of Self-Assessment Questions.) Thus, we have the remarkable phenomenon that analytic continuation around a complete chain does not necessarily return us to the same function element.

In general, suppose that we have a chain

$$\{(f_1, R_1), (f_2, R_2), \ldots, (f_n, R_n)\}$$

with $n > 2$ and such that $R_1 \cap R_n \neq \varnothing$; then there is no guarantee that $f_1 = f_n$ on $R_1 \cap R_n$. (Fig. 17 illustrates such a chain with six elements.)

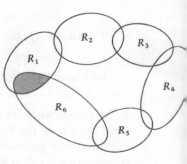

Fig. 17

Luckily, there is a simple criterion for determining whether such behaviour occurs or not, but we require the following definition before formulating it. The following discussion will not help us over our difficulty with the logarithm function, but we shall return to that shortly.

Definition

> A function element (f_1, R_1) is said to be **continued analytically along the polygonal line** $[z_0, z_1, \ldots, z_n]$ if there is a chain $\{(f_1, R_1), (f_2, R_2), \ldots, (f_n, R_n)\}$ such that $[z_k, z_{k+1}] \in R_{k+1}, k = 0, 1, \ldots, n - 1$.

This definition is illustrated in Fig. 18.

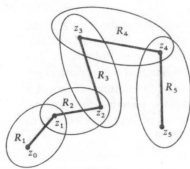

Fig. 18

We can now state the previously mentioned result.

Theorem 8 (The Monodromy Theorem)

Let R be a simply-connected region and let (f_1, R_1) be a function element such that $R_1 \subset R$. If (f_1, R_1) can be continued analytically along every polygonal line in R then there exists a direct analytic continuation (f, R) of (f_1, R_1).

We omit the proof since it is a little tedious and not related to our main line of discussion. (See Levinson and Redheffer, *Complex Variables*, Holden-Day, 1970, for a proof.)

Notice that if the conditions of the theorem are satisfied, then we can find a function element with the largest domain in R, namely (f, R), and once again the process of analytic continuation has served its purpose and becomes redundant.

The following corollary to the Monodromy Theorem is also useful.

Corollary

Let R be a simply-connected region. If (f_1, R_1) is a function element such that $R_1 \subset R$, if (f_1, R_1) can be continued analytically along every polygonal path in R, and if $\{(f_1, R_1), (f_2, R_2), \ldots, (f_n, R_1)\}$ is any complete chain in R, then $f_1 = f_n$.

The corollary follows immediately from the theorem since the function element (f, R) is a direct analytic continuation of both (f_1, R_1) and (f_n, R_1).

Self-Assessment Questions

6. In the notation of page 88, show that $(z \longrightarrow \mathrm{Log}\, z + 2\pi i, P_0)$ is a direct analytic continuation of (L, P_1).

7. In the Monodromy Theorem choose (f_1, R_1) to be (Log, P_0) and choose the region R to be $\mathbf{C} - \{0\}$. Can we conclude that there is a direct analytic continuation $(f, \mathbf{C} - \{0\})$ of (Log, P_0)?

Solutions

6.   We have $P_0 \cap P_1 = \{z : \text{Im } z \neq 0\}$. In the lower half-plane

$$L(z) = \log |z| + 2\pi i + i \text{ Arg } z = \text{Log } z + 2\pi i.$$

In the upper half-plane $L(z) = \text{Log } z$.

7.   No, we cannot. The Monodromy requires that R is *simply-connected*, and $\mathbf{C} - \{0\}$ is not simply-connected.

Summary

We have used the concept of analytic continuation to define the term analytic multi-function. If a function element (or branch) satisfies a functional relationship then so do all other function elements of the analytic multi-function. The Monodromy Theorem gives a condition which ensures that a complete chain in a simply-connected region does not start and finish with different function elements. We have also seen that the logarithm does not contain a function element with largest domain, and that a complete chain of function elements of the logarithm can start and finish with different elements.

11.6 RIEMANN SURFACES

A proper study of Riemann surfaces is too advanced for an introductory course in Complex Analysis, but we have included the following intuitive discussion in the hope that it will give you some further insight into the behaviour of complex functions.

There are two possible approaches to the logarithm function. We saw in *Units 3* and *9* that if we restrict ourselves to the plane we must introduce a cut. We now take a *different* point of view and define the logarithm on a surface. (There are instructions for constructing this surface on page 120.)

You will recall that for the principal branch of the logarithm

$$\text{Log } z = \log|z| + i \text{ Arg } z,$$

where $\text{Arg } z \in (-\pi, \pi)$. The analytic multi-function which we call the logarithm is obtained from this branch by the process of analytic continuation, and if (L, R) is any function element of the logarithm, we have, from the permanence of functional relationships,

$$\exp(L(z)) = z \text{ (and } L'(z) = 1/z), \text{ for } z \in R.$$

Thus, $L(z)$ is a logarithm of z and

$$L(z) = \log|z| + i\theta(z)$$

where $\theta(z)$ is *an* argument of z. Our only difficulty lies in deciding which argument to choose. If (f, R) and (g, S) are two function elements of the logarithm, and z_0 is a point of $R \cap S$, we may have $f(z_0) \neq g(z_0)$ since, as we have seen (page 88), it may be appropriate to choose different values for the argument of z_0 in R and S.

Analytic Continuation along a Contour

There is a relatively simple solution to our problem if we are anxious to assign a value for the logarithm to a point lying on an arc, for we may use the continuous logarithm defined in Section 9.5 of *Unit 9*. We suppose that $\gamma:[a, b] \longrightarrow \mathbf{C}$ is a smooth closed arc, and let

$$l(x) = \int_a^x \frac{\gamma'(t)}{\gamma(t)} dt, \quad x \in [a, b].$$

Then, for some constant λ, $L(x) = l(x) + \lambda$, and $L(x)$ is called a *continuous logarithm* of $\gamma(x)$ for all $x \in [a, b]$.

If Γ is the path of γ from $\gamma(a)$ to $\gamma(x)$ then, putting $w = \gamma(t)$, $\int_a^x \frac{\gamma'(t)}{\gamma(t)} dt$ becomes $\int_\Gamma \frac{dw}{w}$. This suggests that to keep track of "log z" as z moves along a contour Γ we could use the fact that the derivative of every branch of the logarithm is $z \longrightarrow 1/z$ (from the permanence of functional relationships), and evaluate $\int_{z_0}^z \frac{dw}{w}$ (where z_0 is our starting point, and the integral is interpreted as the integral along the contour Γ from z_0 to z). It is not difficult to verify that this process is compatible with the analytic continuation of the logarithm, and we give an outline of the discussion needed.

We can define **analytic continuation along a contour** Γ as follows. We cover Γ with a finite number of open discs D_k, $k = 1, 2, \ldots, n$, of equal radius, such that the centre of D_{k+1} lies in D_k and on Γ, just as we did for a line segment in the proof of Theorem 15 of *Unit 6*.

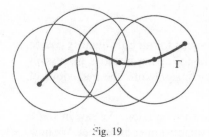

Fig. 19

The discs are also chosen such that, given a function element (f_1, D_1), there is a chain

$$\{(f_1, D_1), (f_2, D_2), \ldots, (f_n, D_n)\}.$$

Consider now the particular case of the logarithm, so that (f_1, D_1) is a branch of the logarithm. We can dissect the contour Γ into contours Γ_k, $k = 1, 2, \ldots, n$, such that $\Gamma_k \subset D_k$ for each k, and $\Gamma = \Gamma_1 + \Gamma_2 + \cdots + \Gamma_n$.

Fig. 20

Then, since $f_k'(z) = 1/z$ for $z \in D_k$, we have

$$\int_{\Gamma_k} \frac{dw}{w} = f_k(z_k) - f_k(z_{k-1}), \quad k = 1, 2, \ldots, n,$$

where z_k and z_{k-1} are the endpoints of Γ_k. We may now add these together (and use the fact that $f_{k+1}(z_k) = f_k(z_k)$) to obtain

$$\int_{\Gamma} \frac{dw}{w} = f_n(b) - f_1(a),$$

where $b = z_n$ and $a = z_0$ are the endpoints of Γ.

Many authors allow a certain abuse of notation and denote *any one* of the branches of the logarithm by log. Using this notation, we may write

$$\int_{\Gamma} \frac{dw}{w} = \log b - \log a.$$

We must, of course, remember that log may denote different branches of the logarithm in the expression on the right, but the analytic continuation along the contour Γ tells us which branch to choose. (We shall shortly define a function which we denote by log, and which is consistent with this notation.)

In practice, the process of evaluating $\int_{\Gamma} \frac{dw}{w}$ is usually quite straightforward. We simply write

$$\int_{\Gamma} \frac{dw}{w} = \log z|_{\Gamma},$$

where $f(z)|_{\Gamma}$ indicates the appropriate change in the function f (here $f = \log$) as we proceed along Γ, from a to b.

But

$$\log z|_{\Gamma} = \log \left| \frac{b}{a} \right| + A(z)|_{\Gamma},$$

where $A(z)|_{\Gamma}$ is the appropriate change in the continuous argument of Γ, and can usually be determined from a diagram (see Section 9.5 of *Unit 5*).

The Riemann Surface for the Logarithm

We have seen that it is impossible to find a function element of the logarithm with largest domain, and so we have been unable to replace this analytic multi-function by one of its function elements, as we could, for example, with $z \longrightarrow 1/(1-z)$.

However, there is a solution to this difficulty which is really rather clever, and it is due to the great German mathematician Riemann.*

Put simply, the idea is this. The difficulty that we are experiencing with the logarithm arises because we can obtain two different function elements (L_1, R_1) and (L_2, R_2) with $R_1 = R_2$. But suppose that we refuse to accept the fact that R_1 and R_2 are identical sets, and insist that a point $z_0 \in R_1$ is somehow distinct from a point $z_0 \in R_2$. It would then be conceivable that the argument of z_0 takes appropriate, distinct, values in R_1 and R_2. In order to prevent R_1 from being identical to R_2, we imagine the regions associated with the function elements (L, R) of the logarithm drawn on a surface, and if $L_1 \neq L_2$ we say that R_1 and R_2 lie on different parts of this surface.

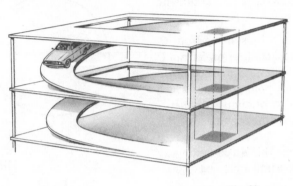 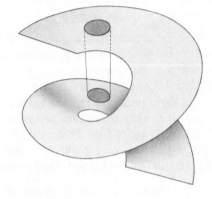

Fig. 21

To give a modern analogy, imagine the complex plane to be the view we would get if we looked vertically down on a multi-storey car park made of glass. Points at different levels may lie on the same vertical line and so correspond to the same complex number. Now imagine the regions corresponding to each of the function elements drawn on the various levels, and on the ramps. As we carry out the process of analytic continuation we may perhaps join regions together which all lie on the second level, in which case a complete chain will take us back to the same function value. On the other hand, some of the regions of a chain may lie on one or more of the ramps, in which case we may well end up vertically above or below our starting point and we get a new function value.

A surface of this kind is called a **Riemann surface**. An appropriate surface for the logarithm is an infinite spiral surface S, part of which is illustrated in Fig. 22. (The adjective "infinite" implies that the surface S extends without bound vertically, both up and down, and laterally.) We can now define **the logarithm** in such a way that it is a function with domain S. For any point z_0 on S we choose a neighbourhood (on S), R say, of z_0 and then find the corresponding function element (L, R) of the logarithm. (We have given two different meanings to R in the above sentence. The second one, as in (L, R), is a region on the complex plane, and is obtained from the first R (which is on S) by projecting R (on S) vertically onto the complex plane.) The **value of the logarithm function at** z_0 is then defined to be $L(z_0)$. If we denote this function by the much used symbol log, then $\log : S \longrightarrow \mathbf{C}$, and $\log z = \log |z| + i\theta(z)$, where $\theta(z)$ is the argument of z determined by the position of z on the Riemann surface.

Fig. 22   Infinite spiral for the logarithm

---

* *Georg Friedrich Bernhard Riemann* (1826–1866) studied at Göttingen with Gauss and later with Dirichlet and Jacobi in Berlin. One of the greatest mathematicians of all time, he gave his name to the Riemann integral, the Cauchy–Riemann equations, Riemann surfaces and the Riemann zeta function. Throughout his life he suffered from poor health, and died at the age of 39.

It is important for this interpretation of the logarithm that we characterize the points of the domain of log, not only as complex numbers but also by their position on the Riemann surface. This characterization is simply achieved if we put $S = \{(r, \theta) : r > 0, \theta \in \mathbf{R}\}$; then the point (r, θ) on the surface corresponds to the complex number $re^{i\theta}$, but the correspondence between complex numbers and points of S is not one-one*. For example, the complex number $1 + i$ corresponds to infinitely many points of S, namely $\left(\sqrt{2}, \dfrac{\pi}{4} + 2n\pi\right)$ for $n = 0, \pm 1$,

$\pm 2, \ldots$, so that $\log(1 + i)$ is not uniquely determined, whereas $\log\left(\sqrt{2}, \dfrac{\pi}{4}\right) =$

$\log \sqrt{2} + i\dfrac{\pi}{4}$. (Note that in this last equation, log on the right-hand side is the real function, and that, strictly speaking $\log(1 + i)$ is meaningless since $1 + i$ is *not* an element of S, the domain of log.) As a device for easy calculation, we can interpret elements (r, θ) of S as complex numbers in polar form $(re^{i\theta})$, in which case we have

$$\log(\sqrt{2}, \pi/4) = \log(\sqrt{2}\,e^{i\pi/4}) = \log \sqrt{2} + i\pi/4.$$

With this convention, real positive numbers are assumed to have zero argument, so that, for example, $1 = e^{i0}$ and $\log(e^{i0}) = \log 1 + i0 = 0$.

We can determine the argument of a given point P on S by starting from Q, some point of S on the positive real axis, and imagining a contour on S from Q to P (any contour will do). The upper part of Fig. 23 shows S marked off in "sectors" of angle $\pi/4$. The portion of S shown shows only points (r, θ) for which $r \leqslant 1$. Notice that the argument of P is the value of the "continuous argument" along the contour at P, since Q has zero argument. Thus, the argument of P is $2\pi + 3\pi/4$. The lower part of Fig. 23 shows how the contour would look if viewed from above: the feint axes provide the orientation. This is a common way of representing contours on S. (See Self-Assessment Question 1 below.)

There are two advantages to this interpretation of the logarithm. We are able to replace the rather daunting notion of an analytic multi-function by a function, albeit with a Riemann surface as its domain, and, moreover, we are able to unify in one concept the various aspects of the "logarithm" introduced so far. The "branch of the logarithm" of *Unit 3*, "the continuous logarithm on an arc" of *Unit 9*, and the analytic multi-function of this unit.

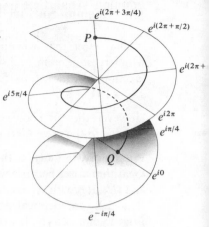

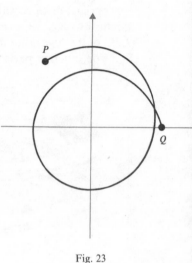

Fig. 23

---

*See Section 1.3 of *Unit 1, Complex Numbers*.

Complete Analytic Functions

Any analytic multi-function can be interpreted as a function with an appropriate Riemann surface as its domain, but in general the surface and its characterization will be rather complicated, and a general discussion is beyond the scope of this course.

This interpretation renders the term "analytic multi-function" redundant, and we prefer the term **complete analytic function** for a function defined on its Riemann surface. For example, log is the complete analytic function defined on the spiral surface S. This extension of a function element by the process of analytic continuation to the corresponding analytic multi-function, and thence to the complete analytic function, is the culmination of our search for the natural domain of a function element. The natural domain of a function is its Riemann surface.

Since locally on the Riemann surface a complete analytic function is identical with one of its function elements (with a suitable interpretation of the domain), we may apply the theory of analytic functions to complete analytic functions, provided that we interpret the regions and curves to lie on the appropriate Riemann surface. (See Problems 2 and 4 of Section 11.7.) In particular, Cauchy's Theorem takes the following form.

Theorem 9   (Cauchy's Theorem for a Complete Analytic Function)

Let R be a simply-connected region on the Riemann surface of a complete analytic function f, and let Γ be a closed contour in R. Then, $\displaystyle\int_\Gamma f = 0$.

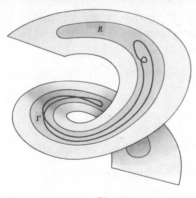

Fig. 24

The adaptation of the notion of simple-connectedness to sets on the Riemann surface is clear, and the proof follows from the fact that f can be decomposed into a chain of function elements for which the union of corresponding regions contains Γ.

Branch Points

The origin has special significance for the "logarithm". For example, in *Unit 9* (page 77, Section 9.3) we showed that there is a branch of the logarithm on every simply-connected region *not containing* 0. This significance becomes even more apparent when we consider the Riemann surface S for the logarithm function, for it is the paths on S which "go round" the origin which take us onto a different level of the surface. Points of this kind are known as *branch points* and are defined as follows.

Definition

> Let z_0 be a point of **C**. If, for a given analytic multi-function f, every neighbourhood of z_0 contains a complete chain $\{(f_1, R_1), (f_2, R_2), \ldots, (f_n, R_1)\}$ of function elements of f such that $f_1 \neq f_n$, then z_0 is said to be a **branch point** of f.

The Power Functions

In *Unit 3* we defined a branch of a power function to be a function of the form $z \longrightarrow \exp(\alpha L_R(z))$, where α is any complex number and L_R is the branch of the logarithm associated with the fundamental region R. We can now extend this definition as follows.

Definition

> Let α be a complex number and let (L, R) be a function element (that is, a branch) of the logarithm. The function $z \longrightarrow \exp(\alpha L(z))$, $z \in R$, is called a **power function element**. The corresponding complete analytic function $z \longrightarrow \exp(\alpha \log z)$ with domain S (the Riemann surface for log) is called a **complete analytic power function**. (For convenience we shall often abbreviate "complete analytic power function" to "power function".)

We will abbreviate $z \longrightarrow \exp(\alpha \log z)$ to $z \longrightarrow z^\alpha$, and the context will make it clear if we intend a branch of the complete analytic function or the complete analytic function itself. (In many contexts either interpretation is equally valid.)

Summary

By introducing the idea of analytic continuation along a contour, we were able to express $\int_\Gamma \frac{dw}{w}$ (which is related to the continuous logarithm) in terms of the continuous argument.

By adopting a different approach to the logarithm we were led to consider Riemann surfaces. The Riemann surface for the logarithm is an infinite spiral surface.

The Riemann surface is the "natural" domain for a function, and a function defined on its Riemann surface is called a complete analytic function. Provided we interpret regions and curves to lie on an appropriate Riemann surface, we may apply the theory of analytic functions to complete analytic functions.

We finished with a brief discussion of branch points and power functions. The infinite spiral surface for the logarithm is also a suitable Riemann surface for power functions.

Self-Assessment Questions

1. In each of the figures (i) and (ii), the point $z_0 = e^{i0}$ is at $(1, 0)$ on the Riemann surface S for log. Calculate $\log z_1$ in each case, where z_1 and the contours are on S.

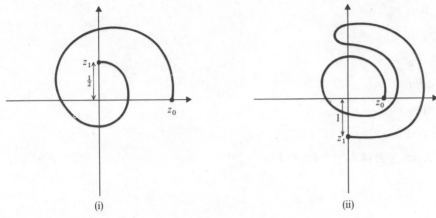

(i) (ii)

Fig. 25

2. Find a suitable Riemann surface for $z \longrightarrow \log(1 + z)$.

3. What are the branch points of

   (i) $z \longrightarrow \log(1 + z)$,

   (ii) $z \longrightarrow \log z + \log(1 + z)$?

4. Calculate (i) 1^i, (ii) $(e^{2\pi i})^i$, (iii) $4^{1/2}$.

Solutions

1. (i) The point z_1 is $\left(\dfrac{1}{2}, \dfrac{5\pi}{2}\right)$ on the Riemann surface, so that

$$\log z_1 = \log\left|\frac{1}{2}\right| + \frac{5\pi i}{2} = -\log 2 + \frac{5\pi i}{2}.$$

   (ii) The point z_1 is $\left(1, \dfrac{3\pi}{2}\right)$ on the Riemann surface, so that

$$\log z_1 = \log|1| + \frac{3\pi i}{2} = \frac{3\pi i}{2}.$$

(If you do construct the Riemann surface S in paper, try to draw the contours in Fig. 25 on your surface.)

2. A suitable Riemann surface is an infinite spiral surface centred on the point -1 of $\mathbf{C}$. This surface can be characterized as the set $\{(r, \phi); r > 0, \phi \in \mathbf{R}\}$ where each of the points $(r, \phi + 2k\pi i)$, $k = 0, \pm 1, \pm 2, \ldots$, corresponds to $1 + z$. In other words, $1 + z$ is written in polar form $re^{i\phi}$ and ϕ determines the position of the point on the Riemann surface.

3. (i) Consideration of the branches of $\log(1 + z)$ shows that $z \longrightarrow \log(1 + z)$ has a branch point at -1. We can also show that there are no other branch points; for, suppose that z_0 is any point of $\mathbf{C}$ other than -1. Let R be a simply-connected region containing z_0 but not -1 (Fig. 26); then there is a branch (f, R) of $z \longrightarrow \log(1 + z)$. If $\{(f_1, R_1), (f_2, R_2), \ldots, (f_n, R_1)\}$ is any complete chain contained in R then, since (f, R) is a direct analytic continuation of both (f_1, R_1) and (f_n, R_1), it follows that $f_1 = f_n$.

   (ii) The branch points of $z \longrightarrow \log z + \log(1 + z)$ are 0 and -1. A similar argument to that above will show that these are the only branch points.

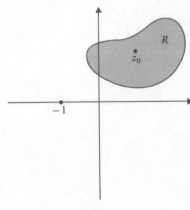

Fig. 26

4.   (i)   According to our convention, $1 = e^{i0}$ corresponds to the point $(1, 0)$ of the Riemann surface S, so that

$$1^i = \exp(i \log e^{i0})$$
$$= \exp(i(\log 1 + i0)) = 1.$$

(ii) $(e^{2\pi i})^i = \exp(i \log e^{2\pi i})$

$$= \exp(i \cdot 2\pi i)$$
$$= \exp(-2\pi) = e^{-2\pi}.$$

(iii) $4^{1/2} = \exp(\tfrac{1}{2} \log 4e^{i0})$

$$= \exp(\tfrac{1}{2} \log 4)$$
$$= \exp(\log 2) = 2.$$

(Note that $4^{1/2} = 2$, not ± 2; $(4e^{2\pi i})^{1/2} = -2$.)

11.7 PROBLEMS

1. Find the value of the power function $z \longrightarrow z^i$ at (i) 3, (ii) $e^{i\pi/2}$.

2. In Fig. 27, $\Gamma_1, \Gamma_2, \Gamma_3$ and Γ_4 form a simple-closed contour lying on S, the Riemann surface for log; Γ_1 and Γ_3 each correspond to the interval $[1, 4]$ of the real line. Is it true that

$$\int_{\Gamma_1} z^{1/2}\, dz = -\int_{\Gamma_3} z^{1/2}\, dz?$$

3. Let (f_0, R_0) be the function element defined by

$$f_0(z) = \sum_{k=0}^{\infty} \left(1 + \frac{1}{1!} + \frac{1}{2!} + \cdots + \frac{1}{k!}\right) z^k,$$

where $0! = 1$, and $R_0 = \{z : |z| < 1\}$. Express $f_0(z)$ in terms of e^z, and hence, for any complete chain of function elements $\{(f_0, R_0), (f_1, R_1), \ldots, (f_n, R_0)\}$, show that

$$f_0 = f_n.$$

4. Let a be a real constant such that $a > 1$, and Γ be the circular contour $\{(a, \theta) : 0 \leqslant \theta \leqslant 2\pi\}$ lying on the Riemann surface for log (Fig. 28). Show that

$$\left| \int_{\Gamma} \frac{z^{1/2}}{1 + z}\, dz \right| < \frac{2\pi a \sqrt{a}}{a - 1}.$$

5. If z_0 is a branch point of an analytic multi-function f, is it necessarily true that f' has a branch point at z_0?

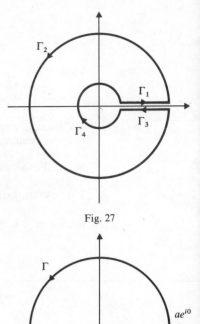

Fig. 27

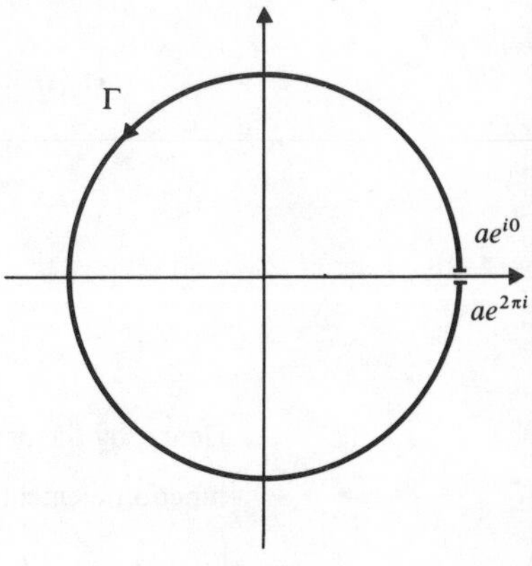

Fig. 28

Solutions

1. (i) Adopting the convention that 3 is the point $(3, 0)$ of the Riemann surface for log, we have

$$3^i = \exp(i \log 3e^{i0})$$

$$= \exp(i \log 3).$$

   (ii) $(e^{i\pi/2})^i = \exp(i \log e^{i\pi/2})$

$$= \exp\left(i \cdot i\frac{\pi}{2}\right) = \exp\left(-\frac{\pi}{2}\right).$$

2. Let us suppose that on Γ_1 (Fig. 27) the points of S are of the form $(r, 2k\pi)$ for some integer k; then, if we put $z = re^{2k\pi i}$ on Γ_1, we obtain

$$z^{1/2} = \exp(\tfrac{1}{2} \log re^{2k\pi i})$$

$$= \exp(\tfrac{1}{2}(\log r + 2k\pi i))$$

$$= \sqrt{r}\, e^{k\pi i}.$$

Hence,

$$\int_{\Gamma_1} z^{1/2}\, dz = e^{k\pi i} \int_1^4 \sqrt{r}\, dr = e^{k\pi i} \tfrac{2}{3} r^{3/2} \big|_1^4 = \tfrac{14}{3} e^{k\pi i}.$$

Since Γ_3 lies on the surface S, it follows that

$$\Gamma_3 = \{(r, \theta) : 1 \leqslant r \leqslant 4, \theta = 2(k + 1)\pi\};$$

thus on Γ_3, putting $z = re^{2(k+1)\pi i}$, we obtain

$$z^{1/2} = \exp(\tfrac{1}{2} \log re^{2(k+1)\pi i})$$

$$= \exp(\tfrac{1}{2}(\log r + 2(k + 1)\pi i))$$

$$= \sqrt{r}\, e^{(k+1)\pi i}$$

$$= \sqrt{r}\, e^{k\pi i} \cdot e^{\pi i}$$

$$= -\sqrt{r}\, e^{k\pi i}.$$

Hence,

$$-\int_{\Gamma_3} z^{1/2}\, dz = \int_{-\Gamma_3} z^{1/2}\, dz$$

$$= \int_1^4 (-\sqrt{r}\, e^{k\pi i})\, dr = -\tfrac{14}{3} e^{k\pi i}.$$

Thus,

$$\int_{\Gamma_1} z^{1/2}\, dz \neq -\int_{\Gamma_3} z^{1/2}\, dz.$$

The essential point here is that the contour lies on S, and we must choose different function elements to give the values of the complete analytic function $z \longrightarrow z^{1/2}$ on Γ_1 and on Γ_3. In fact, we choose one root of z as the value on Γ_1 and the other root to give the value on Γ_3.

3. The function $f_0(z) = \sum_{k=0}^{\infty} \left(1 + \dfrac{1}{1!} + \dfrac{1}{2!} + \cdots + \dfrac{1}{k!}\right) z^k$ is certainly convergent on R_0, the disc $|z| < 1$, since $1 + \dfrac{1}{1!} + \dfrac{1}{2!} + \cdots + \dfrac{1}{k!} < e$. Now

$$f_0(z) = 1 + \left(1 + \frac{1}{1!}\right) z + \left(1 + \frac{1}{1!} + \frac{1}{2!}\right) z^2 + \left(1 + \frac{1}{1!} + \frac{1}{2!} + \frac{1}{3!}\right) z^3 + \cdots$$

$$= \left(1 + \frac{z}{1!} + \frac{z^2}{2!} + \frac{z^3}{3!} + \cdots\right)$$

$$+ \left[z + \left(1 + \frac{1}{1!}\right) z^2 + \left(1 + \frac{1}{1!} + \frac{1}{2!}\right) z^3 + \cdots\right]$$

$$= e^z + z f_0(z).$$

Hence, by the permanence of functional relationships, we have $f_k(z) = \dfrac{e^z}{1 - z}$ for *every* function element (f_k, R_k) of the complete chain $\{(f_0, R_0), (f_1, R_1), \ldots, (f_n, R_0)\}$. Thus $\left(z \longrightarrow \dfrac{e^z}{1 - z}, \mathbf{C} - \{1\}\right)$ is a direct analytic continuation of *every* function element; in particular, this is true of (f_0, R_0) and (f_n, R_0), and therefore $f_0 = f_n$.

4. Putting $z = a e^{i\theta}$, we obtain

$$z^{1/2} = \exp(\tfrac{1}{2} \log a e^{i\theta})$$

$$= \exp(\tfrac{1}{2}(\log a + i\theta))$$

$$= \sqrt{a}\, e^{i\theta/2}.$$

Hence, $|z^{1/2}| = \sqrt{a}$, if $z \in \Gamma$. Also,

$$|z + 1| \geqslant \big||z| - 1\big| = a - 1, \quad \text{if } a > 1.$$

Therefore, from the Estimation Theorem, since the length of Γ is $2\pi a$,

$$\left|\int_{\Gamma} \frac{z^{1/2}}{1 + z}\, dz\right| < \frac{2\pi a \sqrt{a}}{a - 1}.$$

5. The answer is "No". For example, log has a branch point at 0 but $z \longrightarrow 1/z$ has not. (To show that $z \longrightarrow 1/z$ does not have a branch point at 0, consider the function element $(z \longrightarrow 1/z, \mathbf{C} - \{0\})$ and apply the argument of Problem 3.)

11.8 APPLICATIONS

In the final reading section of this unit we shall discuss several applications of the theory which we developed, some of which will be useful for the last three units of the course. Specifically, the gamma function (pages 107 to 110) is required in *Units 14* and *15*, and the Schwarz Reflection Principle (page 113) is required in *Unit 14* (and implicitly in *Unit 16*).

Rouché's Theorem and the Principle of the Argument

In *Unit 10* we proved the Principle of the Argument. We reproduce it here for convenience and justify its name later.

The Principle of the Argument

Let f be a function, meromorphic on a simply-connected bounded region R, and let Γ be a simple-closed contour contained in R. Assume further that f has no zeros or poles *on* Γ. If N is the number of zeros and P is the number of poles of f *inside* Γ, each counted according to its order, then

$$N - P = \frac{1}{2\pi i} \int_\Gamma \frac{f'}{f}.$$

As a corollary, we were able to deduce that

$$\mathrm{Wnd}(f(\Gamma), 0) = N - P,$$

which means that the excess of zeros over poles of a given function inside a closed contour can be determined by counting the number of times the image $f(\Gamma)$ of Γ winds round the origin.

We now show how the above integral can be related to the change in argument of $f(z)$ as z moves along Γ. We can argue as follows, using the substitution $w = f(z)$.

$$\int_\Gamma \frac{f'(z)}{f(z)}\, dz = \int_{f(\Gamma)} \frac{dw}{w} = \log|w|\big|_{f(\Gamma)} + iA(w)\big|_{f(\Gamma)}$$

$$= iA(w)\big|_{f(\Gamma)}, \text{ since } \log|w|\big|_{f(\Gamma)} \text{ is zero, } \Gamma \text{ being closed,}$$

$$= iA(f(z))\big|_\Gamma. \tag{1}$$

We delayed the discussion of the Principle of the Argument because we can obtain this result as follows. The complete analytic function $z \longrightarrow \log f(z)$ has derived function $z \longrightarrow \dfrac{f'(z)}{f(z)}$ so that, applying the Fundamental Theorem for contours on the Riemann surface for $z \longrightarrow \log f(z)$, we obtain

$$\int_\Gamma \frac{f'(z)}{f(z)}\, dz = \log f(z)\big|_\Gamma$$

$$= \log|f(z)|\big|_\Gamma + iA(f(z))\big|_\Gamma$$

$$= iA(f(z))\big|_\Gamma.$$

Theorem 10 (Rouché's Theorem)

Let f and g be analytic on a simply-connected region R and let Γ be a simple-closed contour in R. Let $|f(z)| > |g(z)|$ for each $z \in \Gamma$. Then f and $f + g$ have the same number of zeros inside Γ (each counted according to its order).

Proof

Let $\phi = g/f$, so that $|\phi| < 1$ on Γ and let $N(f)$ and $N(f + g)$ be the number of zeros of f and $f + g$ inside Γ, respectively. Then

$$N(f + g) - N(f) = \frac{1}{2\pi i}\int_\Gamma \frac{f' + g'}{f + g} - \frac{1}{2\pi i}\int_\Gamma \frac{f'}{f}, \quad \text{by the Principle of the Argument,}$$

$$= \frac{1}{2\pi i}\int_\Gamma \frac{fg' - f'g}{f(f + g)}$$

$$= \frac{1}{2\pi i}\int_\Gamma \frac{\phi'}{1 + \phi}, \quad \text{since } \phi = g/f.$$

It remains to show that $\dfrac{1}{2\pi i}\displaystyle\int_\Gamma \frac{\phi'}{1 + \phi} = 0$; but this is precisely the result which we established in Problem 3 of Section 11.4. ∎

Example 1

Show that if $|a| > e$, then there is no complex number z such that $|z| < 1$ and $e^z = az^3$.

Solution

Let $f(z) = e^z$ and $g(z) = -az^3$. On the circle $\Gamma = \{z : |z| = 1\}$ we have $|g(z)| = |a|$. Also $|e^z| = e^x \leqslant e$ if $z \in \Gamma$. Then, if $|a| > e$, the conditions of Rouché's Theorem are satisfied, and we conclude that $z \longrightarrow e^z$ and $z \longrightarrow e^z - az^3$ have the same number of zeros inside Γ. But e^z is never zero, and this establishes the required result.

Example 2 (The Fundamental Theorem of Algebra)

Let p be a polynomial of degree at least 1. Then $p(z) = 0$ for at least one z. Prove this result.

Solution

Let $f(z) = a_n z^n$, $q(z) = a_0 + a_1 z + \cdots + a_{n-1} z^{n-1}$, and $p = q + f$; then, if $|z| = R$,

$$|q(z)| \leqslant |a_0| + |a_1|R + \cdots + |a_{n-1}|R^{n-1},$$

and

$$|f(z)| = |a_n|R^n.$$

But

$$\lim_{R \to \infty} \frac{|a_0| + |a_1|R + \cdots + |a_{n-1}|R^{n-1}}{|a_n|R^n} = 0,$$

so we may choose R so large that

$$|q(z)| < |f(z)| \quad \text{if } |z| = R.$$

But $z \longrightarrow a_n z^n$ has a zero in the disc $|z| < R$ (in fact, it has n), and so $p = q + f$ has a zero in the disc $|z| < R$, by Rouché's Theorem. (Note that the Fundamental Theorem of Algebra was also proved as Theorem 11 of *Unit 5* by means of Liouville's Theorem.)

Self-Assessment Questions

1. In the proof of Rouché's Theorem, we used the Principle of the Argument to deduce that $N = \dfrac{1}{2\pi i}\displaystyle\int_\Gamma f'/f$, but the Principle requires that f has no zeros on Γ. How do we know that f has no zeros on Γ?

2. Let Γ be the closed semicircular contour of radius $r(>3)$ shown in Fig. 29. By applying Rouché's Theorem with $f(z) = z + 2$ and $g(z) = -e^z$, show that the function $z \longrightarrow z + 2 - e^z$ has exactly one zero inside Γ. Deduce that the equation $e^z = z + 2$ has exactly one root in the half-plane $\{z : \operatorname{Re} z < 0\}$.

3. Consider the equation $z^4 + 5z + 1 = 0$.

   (i) Use Rouché's Theorem with $f(z) = 5z$, $g(z) = z^4 + 1$, to find the number of roots of the equation inside the circle $\{z : |z| = 1\}$.

   (ii) Use Rouché's Theorem with $f(z) = z^4$, $g(z) = 5z + 1$, to find the number of roots of the equation inside the circle $\{z : |z| = 2\}$.

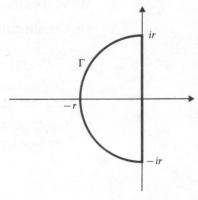

Fig. 29

Solutions

1. In the statement of Rouché's Theorem, we require $|f(z)| > |g(z)|$ for each $z \in \Gamma$, and therefore $|f(z)| \neq 0$ on Γ.

2. If $r > 3$, then $|f| > |g|$ on the contour Γ, since

$$|g(z)| = e^{\operatorname{Re} z} \leqslant 1 < |z + 2| = |f(z)|.$$

   It follows from Rouché's Theorem that the number of zeros of $z \longrightarrow z + 2 - e^z$ inside Γ is the same as the number of zeros of $z \longrightarrow z + 2$ inside Γ, which is 1. The last part follows since the first part is true whatever the value of $r(>3)$.

3. (i) On the circle $\{z : |z| = 1\}$, we have

$$|g(z)| \leqslant |z|^4 + 1 = 2 < |5z| = |f(z)|.$$

   Hence, by Rouché's Theorem, the number of roots of the equation inside this circle is equal to the number of zeros of $z \longrightarrow 5z$ inside the circle, which is *one*.

   (ii) On the circle $\{z : |z| = 2\}$, we have

$$|g(z)| \leqslant 5|z| + 1 = 11 < 16 = |z^4| = |f(z)|.$$

   Hence, by Rouché's Theorem, the number of roots of the equation inside this circle is equal to the number of zeros of $z \longrightarrow z^4$ inside the circle, which is *four*.

Problems

1. Use the method of Self-assessment Question 3 on page 103 to prove that the equation $2z^5 + 8z - 1 = 0$ has exactly four roots in the annulus $\{z : 1 < |z| < 2\}$.

    Where does the other root lie?

2. Give an alternative proof of Rouché's Theorem, as follows:

    (i)   If $F(z) = \dfrac{f(z) + g(z)}{f(z)}$, show that $|F(z) - 1| < 1$ for all points z of Γ.

    Deduce that Γ lies entirely within the circle centre 1, with radius 1, and that the continuous argument $A(F(z))$ cannot vary by more than π.

    (ii) Apply the Principle of the Argument to F, and deduce that $N = P$ for the function F. Hence deduce Rouché's Theorem.

Solutions

1. Taking Rouché's Theorem with $f(z) = 2z^5, g(z) = 8z - 1$, we see that all five roots of $2z^5 + 8z - 1 = 0$ lie within the circle $C = \{z : |z| = 2\}$, since on C,

    $$|g(z)| \leqslant 8|z| + 1 = 17 < 64 = |f(z)|.$$

    Similarly, taking $f(z) = 8z - 1, g(z) = 2z^5$, we see that there is only one root inside the circle $\{z : |z| = 1\}$.

    There are clearly no roots *on* the circle $\{z : |z| = 1\}$, since $|2z^5| = 2$ and $|8z - 1| > 7$. So there are exactly four roots in the annulus $\{z : 1 < |z| < 2\}$.

    The other root clearly lies inside the unit circle. But we can say a lot more than that, since if $p(z) = 2z^5 + 8z - 1$, then $p(0) < 0$ and $p(1) > 0$. It follows, from the Intermediate Value Theorem (**Spivak**, page 100), that there is a *real* root between 0 and 1. In fact, since $p(\frac{1}{8}) > 0$, one such root lies on the real axis between 0 and $\frac{1}{8}$.

2. (i)   Since $|g| < |f|$ on Γ, it follows that

    $$|F(z) - 1| = \left| \frac{f(z) + g(z)}{f(z)} - 1 \right| = \left| \frac{g(z)}{f(z)} \right| < 1,$$

    for all points z of Γ. The rest of part (i) then follows immediately (see Fig. 30).

    (ii) By the Principle of the Argument and Equation (1) on page 101, we have $N - P = \dfrac{1}{2\pi} A(F(z))|_\Gamma$. But, by part (i), $\left| A(F(z))|_\Gamma \right| < \pi$, and so we must have $A(f(z))|_\Gamma = 0$, since $N - P$ is an integer. Hence $N = P$. But N is the number of zeros of F inside Γ, which is equal to the number of zeros of $f + g$ inside Γ, and P is the number of poles of F inside Γ, which is equal to the number of zeros of f inside Γ. The result follows.

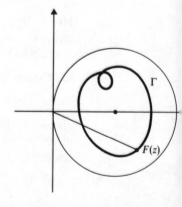

Fig. 30

More Contour Integrals

In Unit 10 we showed how it is possible to evaluate certain real integrals using the Residue Theorem. If we apply the same technique to complete analytic functions then we can determine the value of a much wider class of integrals; but first we have to generalize some of our definitions.

Definition

> The complete analytic function f is said to have a **pole with residue α at a point P** of its Riemann surface if the function elements of f, corresponding to every sufficiently small punctured neighbourhood of P on the surface, have poles with residue α at the complex numbers corresponding to P.

For example, the complete analytic function $z \longrightarrow \dfrac{\log z}{1 - z^2}$ has simple poles at the points $e^{k\pi i}$, $k = 0, \pm 1, \pm 2, \ldots$ on the Riemann surface for log, for these are the points of the surface which correspond to 1 and -1, the zeros of $z \longrightarrow z^2 - 1$.

Theorem 11   (The Residue Theorem for Complete Analytic Functions)

Let R be a simply-connected region on the Riemann surface of a complete analytic function f, and suppose that R contains the points α_j, $j = 1, 2, \ldots, n$ (not necessarily distinct). Let Γ be any simple-closed contour in R that does not pass through any point α_j.

If $\phi(z) = \dfrac{f(z)}{(z - \alpha_1)(z - \alpha_2) \cdots (z - \alpha_n)}$, then

$$\int_{\Gamma} \phi = 2\pi i \sum (\text{residue of } \phi \text{ at } \alpha_j)$$

where the sum extends over those points α_j which lie inside Γ.

We omit the proof, which is again based on decomposing ϕ into function elements, and proceed directly to an example.

Example 3

Use contour integration to evaluate

$$\int_0^\infty \frac{1}{(1 + x)\sqrt{x}}\, dx.$$

Solution

Consider the complete analytic function $f : z \longrightarrow \dfrac{z^{-1/2}}{1 + z}$, which has a branch point at 0. (Remember that $z^{-1/2} = \exp(-\tfrac{1}{2} \log z)$.)

We choose the contour $\Gamma = \Gamma_1 + \Gamma_2 + \Gamma_3 + \Gamma_4$ illustrated in Fig. 31, which we suppose is drawn on the Riemann surface S for log. Notice that this contour is carefully chosen to avoid the branch point at 0, and both Γ_1, and Γ_3 correspond to the interval $[\varepsilon, r]$ of the real axis. However, following our previous convention (page 99), the points of Γ_1, have argument 0 and the points of Γ_3 have argument 2π.

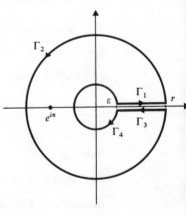

Fig. 31

The function $z \longrightarrow \dfrac{z^{-1/2}}{1+z}$ has a pole at the point $e^{i\pi}$ of S with residue $(e^{i\pi})^{-1/2} = -i$, and this pole is inside Γ if $0 < \varepsilon < 1 < r$; so that, applying the Residue Theorem, we obtain

$$\int_\Gamma \frac{z^{-1/2}}{1+z}\,dz = \int_{\Gamma_1} f + \int_{\Gamma_2} f + \int_{\Gamma_3} f + \int_{\Gamma_4} f$$

$$= \int_\varepsilon^r \frac{dx}{(1+x)\sqrt{x}} + \int_0^{2\pi} \frac{(re^{i\theta})^{-1/2} ire^{i\theta}}{1 + re^{i\theta}}\,d\theta + \int_r^\varepsilon \frac{(e^{2\pi i}x)^{-1/2}}{1+x}\,dx$$

$$+ \int_{2\pi}^0 \frac{(\varepsilon e^{i\theta})^{-1/2} i\varepsilon e^{i\theta}}{1 + \varepsilon e^{i\theta}}\,d\theta$$

$$= 2\pi i \times (\text{residue at } e^{i\pi}) = 2\pi.$$

The first and third integrals do not cancel, but combine to give

$$[1 - (e^{2\pi i})^{-1/2}]\int_\varepsilon^r \frac{dx}{(1+x)\sqrt{x}} = 2\int_\varepsilon^r \frac{dx}{(1+x)\sqrt{x}}.$$

Also, for the second integral,

$$\left| \int_{\Gamma_2} f \right| \leqslant \frac{2\pi r^{1/2}}{r-1}, \quad \text{since } r > 1,$$

so that, $\displaystyle\lim_{r\to\infty} \int_{\Gamma_2} f = 0$.

For the fourth integral,

$$\left| \int_{\Gamma_4} f \right| \leqslant \frac{2\pi \varepsilon^{1/2}}{1-\varepsilon}, \quad \text{since } \varepsilon < 1,$$

so that, $\displaystyle\lim_{\varepsilon\to 0} \int_{\Gamma_4} f = 0$.

Hence, $\displaystyle\int_0^\infty \frac{dx}{(1-x)\sqrt{x}} = \pi$.

It may come as a surprise that the integrals along Γ_1 and Γ_3 do not cancel, but this is a consequence of the fact that Γ_1 and Γ_3 lie on different parts of the Riemann surface. (Problem 2 of Section 11.7 was another illustration of this circumstance.)

This integral may also be evaluated by means of the substitution $x = t^2$, $dx = 2t\,dt$, but the following Self-assessment Question shows the power of contour integration in the general case.

Self-Assessment Question 4

Let a be a number such that $0 < a < 1$. By considering the integral of the complete analytic function $z \longrightarrow \dfrac{z^{a-1}}{1+z}$ along the contour Γ of the previous example, evaluate $\displaystyle\int_0^\infty \frac{x^{a-1}}{1+x}\,dx$.

Solution

The argument is the same as in the example except that z^{1-a} replaces $z^{1/2}$ throughout. The value of the integral is $\dfrac{\pi}{\sin \pi a}$.

Problem 3

Evaluate $\displaystyle\int_0^\infty \frac{\log x}{x^2 + b^2}\,dx$, $b > 0$, by integrating the complete analytic function

$z \longrightarrow \dfrac{\log z}{z^2 + b^2}$ along the contour $\Gamma = \Gamma_1 + \Gamma_2 + \Gamma_3 + \Gamma_4$ shown in Fig. 32.

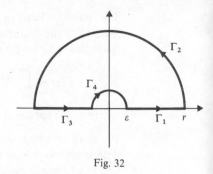

Fig. 32

Solution

Inside the contour Γ, the function $f : z \longrightarrow \dfrac{\log z}{z^2 + b^2}$ has a pole at $be^{i\pi/2}$ with residue

$$\lim_{z \to be^{i\pi/2}} \frac{(z - be^{i\pi/2})\log z}{z^2 + b^2} = \frac{\log b + i\pi/2}{2be^{i\pi/2}} = \frac{\log b + i\pi/2}{2ib}.$$

From the Residue Theorem,

$$\int_\Gamma f = \int_{\Gamma_1} f + \int_{\Gamma_2} f + \int_{\Gamma_3} f + \int_{\Gamma_4} f$$

$$= \int_\varepsilon^r \frac{\log x}{x^2 + b^2}\,dx + \int_0^\pi \frac{\log r + i\theta}{(re^{i\theta})^2 + b^2} ire^{i\theta}\,d\theta + \int_r^\varepsilon \frac{(\log(xe^{i\pi})e^{i\pi}}{(xe^{i\pi})^2 + b^2}\,dx$$

$$+ \int_{-\pi}^0 \frac{\log \varepsilon + i\theta}{(\varepsilon e^{i\theta})^2 + b^2} i\varepsilon e^{i\theta}\,d\theta$$

$$= 2\pi i \ (\text{residue at } be^{i\pi/2}) = \frac{\pi}{b}(\log b + i\pi/2).$$

The first and third integrals combine to give

$$2\int_\varepsilon^r \frac{\log x}{x^2 + b^2}\,dx + \int_\varepsilon^r \frac{i\pi}{x^2 + b^2}\,dx.$$

Also

$$\left| \int_{\Gamma_2} f \right| \leqslant \frac{\pi r(\log r + \pi)}{r^2 - b^2} \quad \text{if } r > b, \text{ so that, } \lim_{r \to \infty} \int_{\Gamma_2} f = 0,$$

and

$$\left| \int_{\Gamma_4} f \right| \leqslant \frac{\pi\varepsilon(\log \varepsilon + \pi)}{b^2 - \varepsilon^2} \quad \text{if } \varepsilon < b, \text{ so that, } \lim_{\varepsilon \to 0^+} \int_{\Gamma_4} f = 0.$$

It follows that, if $\varepsilon < b < r$,

$$2\int_0^\infty \frac{\log x}{x^2 + b^2}\,dx + \int_0^\infty \frac{i\pi}{x^2 + b^2}\,dx = \frac{\pi}{b}(\log b + i\pi/2),$$

so that, equating real parts, we have

$$\int_0^\infty \frac{\log x}{x^2 + b^2}\,dx = \frac{\pi}{2b}\log b.$$

(Notice that equating imaginary parts provides us with a check on our working, since $\displaystyle\int \frac{dx}{x^2 + b^2} = \frac{1}{b}\arctan\frac{x}{b}$.)

The Gamma Function

You may remember that we can define the elementary real functions sin, cos, exp, etc. analytically, and then deduce their properties from their analytic definitions. You may also recall that this is the approach which we adopted in M231 Analysis (**Spivak**, Chapter 15), and you may have wondered why we tackle them this way rather than geometrically. One reason is that this approach is the one which we are forced to adopt with the more advanced functions, and among the most important of these is the *gamma function*.

The gamma function was introduced by Leonhard Euler* who wished to generalize the factorial function by constructing a continuous function which takes the value $n!$ at the natural number n. Euler's **gamma function** is defined by

$$\Gamma(z) = \int_0^\infty t^{z-1}e^{-t}\,dt,$$

where t^{z-1} represents the principal branch of $t \longrightarrow \exp((z-1)\log t)$.

We shall establish some of its simpler properties, and our first step will be to determine on which regions the gamma function is analytic and for which values of z the integral converges.

We wish to use the Weierstrass M-test and so we break the integral up to form an infinite series of functions, but in order to establish the various inequalities we assume that z lies in the half-plane $\{z: \mathrm{Re}\, z > 1\}$.

Let

$$p_n(z) = \int_n^{n+1} t^{z-1}e^{-t}\,dt, \qquad n = 0, 1, 2, \ldots;$$

then we can verify that the conditions of Theorem 5 (page 76) hold (with $H(w, z) = w^{z-1}e^{-w}$), and thus show that each function p_n is analytic on $\{z: \mathrm{Re}\, z > 1\}$:

(i) $t \longrightarrow t^{z-1}e^{-t}$ is continuous on $[n, n+1]$ for each fixed z in $\{z: \mathrm{Re}\, z > 1\}$;

(ii) $z \longrightarrow t^{z-1}e^{-t}$ is analytic on $\{z: \mathrm{Re}\, z > 1\}$ for each fixed $t \in [n, n+1]$.

Also since

$$|t^{z-1}e^{-t}| = t^{x-1}e^{-t}, \quad (z = x + iy),$$

it follows that, if $1 < x \leqslant \alpha$ and $n \leqslant t \leqslant n+1$, then

$$|t^{z-1}e^{-t}| \leqslant (n+1)^{\alpha-1}e^{-n},$$

where α is any number greater than 1.

From Theorem 5 we deduce that p_n is analytic on any bounded region in $\{z: \mathrm{Re}\, z > 1\}$, and therefore p_n is analytic on this half-plane.

In order to apply the Weierstrass M-test, we require an upper estimate for $|p_n(z)|$, and this can easily be obtained from the above inequality; for,

$$|p_n(z)| = \left| \int_n^{n+1} t^{z-1}e^{-t}\,dt \right|$$

$$\leqslant \int_n^{n+1} |t^{z-1}e^{-t}|\,dt$$

$$\leqslant (n+1)^{\alpha-1}e^{-n}, \quad \text{if } 1 < \mathrm{Re}\, z < \alpha.$$

But the series $\sum_{n=0}^{\infty} (n+1)^{\alpha-1}e^{-n}$ is convergent (by the ratio test) and it follows that the series $\sum_{n=0}^{\infty} p_n(z)$ converges uniformly on any closed disc in the infinite strip $\{z: 1 < \mathrm{Re}\, z < \alpha\}$. We deduce that the function ϕ defined by

$$\phi(z) = \sum_{n=0}^{\infty} p_n(z) = \lim_{N \to \infty} \int_0^N t^{z-1}e^{-t}\,dt$$

is analytic on this infinite strip.

---

* *Leonhard Euler* (1707–1783) was one of the greatest mathematicians of all time. He was born in Basle, but spent most of his life in St. Petersburg and Berlin. He made fundamental contributions to all branches of pure mathematics then in existence, as well as writing papers in mechanics, physics and astronomy. For the last fifteen years of his life he was completely blind, and died of apoplexy.

It remains to show that ϕ is, in fact, the gamma function. The point is that in the above expression we have considered only the limit through *integer* values, but the definition of $\Gamma(z)$ allows N to get large through all values. We now have to tidy up this point.

If we let $[h]$ denote the largest integer $\leq h$ then,

$$\left| \int_{[h]}^{h} t^{z-1}e^{-t}\, dt \right| \leq \int_{[h]}^{h} t^{x-1}e^{-t}\, dt.$$

If $t \in [[h], h]$, then $e^{-t} \leq e^{-[h]} \leq e^{-(h-1)}$; so

$$\left| \int_{[h]}^{h} t^{z-1}e^{-t}\, dt \right| \leq h^{x-1}e^{1-h}, \quad \text{if } x > 1,$$

and $\lim\limits_{h \to \infty} h^{x-1}e^{1-h} = 0$.

It follows that

$$\lim_{h \to \infty} \int_0^h t^{z-1}e^{-t}\, dt = \lim_{h \to \infty} \int_0^{[h]} t^{z-1}e^{-t}\, dt.$$

In other words, it does not matter if the limit is taken through integer values or arbitrary values.

We deduce that $\Gamma(z) = \phi(z)$ and hence the gamma function is analytic on the half-plane $\{z : \operatorname{Re} z > 1\}$. (We could have used the result of Example 2 of Section 15.3 instead of the above method; however, the method we employed provides useful practice in other techniques.)

The integral $\Gamma(z) = \int_0^\infty t^{z-1}e^{-t}\, dt$ provides us with a function element of Γ with the half-plane $\{z : \operatorname{Re} z > 1\}$ as its domain, but now we ask whether Γ can be continued analytically onto a larger region.

Integration by parts gives

$$\Gamma(z+1) = \int_0^\infty t^z e^{-t}\, dt = -t^z e^{-t}\Big|_0^\infty + \int_0^\infty z t^{z-1}e^{-t}\, dt$$

$$= z\Gamma(z), \quad \text{provided that } \operatorname{Re} z > 1.$$

From the permanence of functional relationships it follows that $\Gamma(z+1) = z\Gamma(z)$ for *all* function elements of Γ, and this fact can be used to derive a method of analytic continuation.

If we now write $\Gamma(z) = \dfrac{\Gamma(z+1)}{z}$ and notice that $z \longrightarrow \Gamma(z+1)$ is analytic on $\{z : \operatorname{Re}(z+1) > 1\}$ (that is, on $\{z : \operatorname{Re} z > 0\}$), then we can deduce that $z \longrightarrow \Gamma(z)$ is also analytic on $\{z : \operatorname{Re} z > 0\}$.

Notes

(i) We have found an analytic continuation of Γ from $\{z : \operatorname{Re} z > 1\}$ onto $\{z : \operatorname{Re} z > 0\}$. This may come as a slight surprise, but a re-examination of the integral defining $\Gamma(z)$ explains what has happened.

We have $\Gamma(z) = \int_0^\infty t^{z-1}e^{-t}\, dt$, and this integral actually converges for $\operatorname{Re} z > 0$, in the sense that $\lim\limits_{\substack{h \to \infty \\ \varepsilon \to 0^+}} \int_\varepsilon^h t^{z-1}e^{-t}\, dt$ exists. (Notice that we must introduce the limit near zero because of the behaviour of t^{z-1} near $t = 0$.)

To show that this is so, we can use integration by parts to give

$$\int_\varepsilon^h t^{z-1}e^{-t}\, dt = \frac{t^z e^{-t}}{z}\Big|_\varepsilon^h + \frac{1}{z}\int_\varepsilon^h t^z e^{-t}\, dt.$$

The previous discussion shows that $\int_0^\infty t^z e^{-t}\, dt$ exists if $\mathrm{Re}\, z > 0$, so that there is no difficulty with the integral on the right. Also, $\lim_{h \to \infty} h^z e^{-h} = 0$, and $\lim_{\varepsilon \to 0^+} \varepsilon^z e^{-\varepsilon} = 0$ if $\mathrm{Re}\, z > 0$. It follows that $\int_0^\infty t^{z-1} e^{-t}\, dt$ exists if $\mathrm{Re}\, z > 0$, and moreover, it equals $\dfrac{\Gamma(z+1)}{z}$.

(ii) We showed that $\Gamma(z+1) = z\Gamma(z)$, and since $\Gamma(1) = \int_0^\infty e^{-t}\, dt = 1$ and $\Gamma(n+1) = n\Gamma(n)$, it follows that $\Gamma(n+1) = n!$ for each positive integer n.

We now return to the analytic continuation of Γ and again we use the functional equation $\Gamma(z) = \dfrac{\Gamma(z+1)}{z}$. We know that $z \longrightarrow \Gamma(z+1)$ is analytic on $\{z : \mathrm{Re}(z+1) > 0\}$ (that is, on $\{z : \mathrm{Re}\, z > -1\}$), and so we have obtained a direct analytic continuation of Γ onto the half-plane $\{z : \mathrm{Re}\, z > -1\}$ with the origin removed. At the origin, Γ has a simple pole with residue $\Gamma(1) = 1$.

We may now repeat the process and obtain

$$\Gamma(z) = \frac{\Gamma(z+1)}{z} = \frac{\Gamma(z+2)}{z(z+1)},$$

which enables us to continue Γ onto $\{z : \mathrm{Re}\, z > -2,\ z \neq 0,\ z \neq -1\}$. We also see that Γ has a simple pole at -1 with residue $\dfrac{\Gamma(1)}{-1} = -1$. Repeated application of this method produces

$$\Gamma(z) = \frac{\Gamma(z+n+1)}{z(z+1)(z+2)\cdots(z+n)},$$

for each integer n, and therefore Γ is analytic on $\mathbf{C}$ except for simple poles at $0, -1, -2, \ldots$. The residue at $-n$ is

$$\lim_{z \to -n} (z+n)\Gamma(z) = \frac{1}{-n(1-n)(2-n)\cdots(-1)}$$

$$= \frac{(-1)^n}{n!}.$$

Self-Assessment Questions

5. Using the fact that $\int_{-\infty}^\infty e^{-x^2}\, dx = \sqrt{\pi}$, and a substitution in the integral $\int_0^\infty e^{-x^2}\, dx$, evaluate $\Gamma(\tfrac{1}{2})$.

6. Calculate $\Gamma(\tfrac{3}{2})$.

Solutions

5. $\displaystyle \int_{-\infty}^\infty e^{-x^2}\, dx = 2\int_0^\infty e^{-x^2}\, dx$

    $\displaystyle \qquad = \int_0^\infty \frac{e^{-t}}{\sqrt{t}}\, dt, \quad \text{using the substitution } x^2 = t,$

    $\displaystyle \qquad = \Gamma(\tfrac{1}{2}).$

    Thus, $\Gamma(\tfrac{1}{2}) = \sqrt{\pi}$.

6. We have $\Gamma(z+1) = z\Gamma(z)$ and $\Gamma(\tfrac{1}{2}) = \sqrt{\pi}$; so that $\Gamma(\tfrac{3}{2}) = \tfrac{1}{2}\Gamma(\tfrac{1}{2}) = \dfrac{\sqrt{\pi}}{2}$.

Problems

4. Integrate $f : z \longrightarrow z^{a-1} e^{-z}$ along the contour $C = C_1 + C_2 + C_3 + C_4$ shown in Fig. 33, and hence prove that

$$\int_0^\infty x^{a-1} \sin x \, dx = \Gamma(a) \sin \frac{\pi a}{2}, \quad \text{if } 0 < a < 1.$$

This result will be used in *Unit 15, Number Theory.*

(Hint: Use the result

$$\int_0^{\pi/2} e^{-\beta r \sin \theta} \, d\theta < \frac{\pi}{2\beta r},$$

which we established when proving Jordan's Lemma in Section 10.4 of *Unit 10.*)

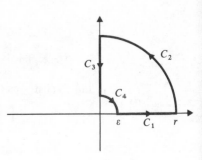

Fig. 33

5. Let f be a real function continuous on the interval $[0, b]$ for every $b > 0$ and suppose that $|f(t)| \leqslant M e^{ct}, t \in [0, \infty)$, for some real constants M and c. Show that the integral

$$F(z) = \int_0^\infty e^{-zt} f(t) \, dt$$

exists and defines an analytic function in the half-plane $\{z : \operatorname{Re} z > c\}$. (Hint: Use the method which established the analyticity of the gamma function.) Show also that

$$F^{(n)}(z) = \int_0^\infty (-t)^n e^{-zt} f(t) \, dt, \quad \text{if } \operatorname{Re} z > c.$$

(The function $F : z \longrightarrow \int_0^\infty e^{-zt} f(t) \, dt$ is called the Laplace transform of f: we shall discuss it in *Unit 14, Laplace Transforms.*)

Solutions

4. Taking the principal branch of the function $z \longrightarrow z^{a-1}$, the function $f : z \longrightarrow z^{a-1} e^{-z}$ is analytic on a region containing the given contour C. From Cauchy's Theorem,

$$\int_C f = 0,$$

where $C = C_1 + C_2 + C_3 + C_4$ (see Fig. 33).

(i) We have $\int_{C_1} f = \int_\varepsilon^r x^{a-1} e^{-x} \, dx$, and therefore

$$\lim_{\substack{\varepsilon \to 0^+ \\ r \to \infty}} \int_{C_1} f = \int_0^\infty x^{a-1} e^{-x} \, dx = \Gamma(a).$$

(ii) On C_2, $z = re^{i\theta}$ and $dz = rie^{i\theta} \, d\theta$, and so

$$\left| \int_{C_2} f \right| = \left| \int_{C_2} z^{a-1} e^{-z} \, dz \right| = \left| \int_0^{\pi/2} r^{a-1} (ie^{i\theta})^{a-1} e^{-r(\cos\theta + i\sin\theta)} rie^{i\theta} \, d\theta \right|$$

$$\leqslant \int_0^{\pi/2} r^a e^{-r\cos\theta} \, d\theta,$$

since $|(ie^{i\theta})^{a-1} e^{-ir\sin\theta}| = 1$.

But now we can use the inequality mentioned in the hint and observe that

$$\int_0^{\pi/2} e^{-r\cos\theta} \, d\theta = \int_0^{\pi/2} e^{-r\sin\theta} \, d\theta,$$

so that

$$\left| \int_{C_2} f \right| \leqslant r^a \cdot \frac{\pi}{2r}$$

$$= \frac{\pi}{2} r^{a-1}.$$

Hence, $\lim_{r \to \infty} \int_{C_2} f = 0$, since $a < 1$.

(iii) Similarly, on C_4, $z = \varepsilon e^{i\theta}$ and $dz = i\varepsilon e^{i\theta}$, and we obtain

$$\left| \int_{C_4} f \right| = \left| \int_0^{\pi/2} \varepsilon^{a-1}(ie^{i\theta})^{a-1} e^{-\varepsilon(\cos\theta + i\sin\theta)} \varepsilon i e^{i\theta} \, d\theta \right|$$

$$\leqslant \int_0^{\pi/2} \varepsilon^a e^{-\varepsilon\cos\theta} \, d\theta$$

$$\leqslant \frac{\pi}{2} \varepsilon^a.$$

Hence, $\lim_{\varepsilon \to 0^+} \int_{C_4} f = 0$, since $a > 0$.

(iv) We have

$$\int_{C_3} f = -\int_\varepsilon^r (iy)^{a-1} e^{-iy} i \, dy$$

$$= -\int_\varepsilon^r y^{a-1} e^{-iy} e^{(i\pi a)/2} \, dy.$$

Combining these results, we obtain

$$\Gamma(a) = \int_0^\infty y^{a-1} e^{-iy} e^{(i\pi a)/2} \, dy,$$

so that

$$e^{-i\pi/2}\Gamma(a) = \int_0^\infty y^{a-1} e^{-iy} \, dy,$$

and equating imaginary parts gives the required result.

5. We give an outline which uses the method which established the analyticity of the gamma function. Let $P_n(z) = \int_n^{n+1} e^{-zt} f(t) \, dt$, and let $c \geqslant 0$; then

$$|P_n(z)| \leqslant e^{-nx} M e^{c(n+1)} = M e^c e^{n(c-x)}, \quad \text{if } z \in [n, n+1].$$

For any $k > c$, we have $|P_n(z)| < M e^c e^{n(c-k)}$, if $x > k$, and $\sum_{n=0}^\infty \frac{M e^c}{e^{n(k-c)}}$ is convergent. From the Weierstrass M-test, it follows that $\sum_{n=0}^\infty P_n(z)$ is uniformly convergent; since each function P_n is entire (from Theorem 5), it then follows that $\lim_{n \to \infty} \int_0^n e^{-zt} f(t) \, dt$ exists. (If $c < 0$, then $|P_n(z)| \leqslant e^{-nx} M e^{-cn}$, if $z \in [n, n+1]$ and the same result follows.) As in the earlier proof, we can now show that $\lim_{k \to \infty} \int_0^k e^{-zt} f(t) \, dt$ exists. We can equally show that

$$F^{(n)}(z) = \lim_{n \to \infty} \int_0^n (-t)^n e^{-zt} f(t) \, dt,$$

using the fact that $\sum_{n=0}^\infty P_n(z)$ is uniformly convergent on any closed disc in the half-plane $\{z : \operatorname{Re} z > c\}$.

(We can use the same method as before to show that $\int_0^\infty (-t)^n e^{-zt} f(t) \, dt$ exists.)

The Schwarz Reflection Principle

The Schwarz* reflection principle provides a particularly simple method of analytic continuation for functions which are real-valued on the real axis.

Theorem 12   (The Schwarz Reflection Principle)

Let R be a region which has a line segment L of the real axis as part of its boundary, and let f be a function analytic on a region which contains $R \cup L$ such that $f(z)$ is real on L. If $\tilde{R} = \{z : \bar{z} \in R\}$, then $(z \longrightarrow \overline{f(\bar{z})}, \tilde{R})$ and (f, R) are function elements of the same complete analytic function.

Proof

The function $z \longrightarrow \overline{f(\bar{z})}$ is analytic on a region containing $\tilde{R} \cup L$. (See Problem 2(ii) of Section 3.3 of *Unit 3*.) Now define a function on $R \cup \tilde{R} \cup L$ by putting

$$\phi(z) = \begin{cases} f(z), & \text{if } z \in R \cup L \\ \overline{f(\bar{z})}, & \text{if } z \in \tilde{R}. \end{cases}$$

Let Δ be any triangle in $R \cup \tilde{R} \cup L$, and consider the following three cases.

(i) $\Delta \subset R$: in which case $\int_{\Delta} \phi = \int_{\Delta} f = 0$, from Cauchy's Theorem.

(ii) $\Delta \subset \tilde{R}$: in which case $\int_{\Delta} \phi = \int_{\Delta} (z \longrightarrow \overline{f(\bar{z})}) = 0$, again from Cauchy's Theorem.

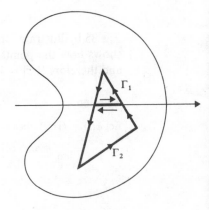

(iii)   In the remaining case we can decompose Δ as shown in Fig. 34 and, since $f(z) = \overline{f(\bar{z})}$ on L, it follows that

$$\int_{\Delta} \phi = \int_{\Gamma_1} \phi + \int_{\Gamma_2} \phi$$

$$= \int_{\Gamma_1} f + \int_{\Gamma_2} (z \longrightarrow \overline{f(\bar{z})})$$

$$= 0, \quad \text{from Cauchy's Theorem.} \quad \blacksquare$$

Fig. 34

Self-Assessment Question 7

Use the Schwarz reflection principle to show that the zeros of an entire function f which is real-valued on the real axis must occur in conjugate pairs.

Solution

By the Schwarz reflection principle, $\overline{f(\bar{z})} = f(z)$, $z \in \mathbf{C}$. If $f(a) = 0$, then $\overline{f(\bar{a})} = 0$.

---

* *Hermann Schwarz* (1843–1921) was a pupil of Weierstrass, and succeeded him as professor in Berlin. He made worthwhile contributions to a wide variety of subjects, including the theory of surfaces, conformal mapping (see *Unit 13*) and potential theory.

Problem 6

If f is differentiable at a point α, show that $z \longrightarrow \overline{f(1/\bar{z})}$ is differentiable at $1/\bar{\alpha}$.

Let R be a region which has an arc L of the unit circle $\{z : |z| = 1\}$ as part of its boundary, and let f be a function which is analytic on a region containing $R \cup L$ such that $f(z)$ is real on L. If $S = \{z : 1/\bar{z} \in R\}$ show that $(z \longrightarrow \overline{f(1/\bar{z})}, S)$ is a function element of the same complete analytic function as $(f, R \cup L)$.

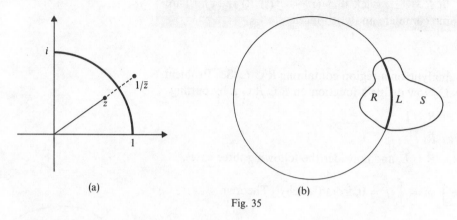

(a)   (b)

Fig. 35

Fig. 35(b) illustrates only the case where R is a subset of the unit circle. Fig. 35(a) shows how the points of S are obtained: the point z shown has modulus 0.8, and therefore $|1/\bar{z}| = 1.25$.

Solution

Let $\phi(z) = \overline{f(1/\bar{z})}$; then

$$\lim_{z \to 1/\bar{\alpha}} \frac{\phi(z) - \phi(1/\bar{\alpha})}{z - 1/\bar{\alpha}} = \lim_{z \to \alpha} \frac{\phi(1/\bar{z}) - \phi(1/\bar{\alpha})}{1/\bar{z} - 1/\bar{\alpha}}$$

$$= \lim_{z \to \alpha} \frac{\overline{f(z)} - \overline{f(\alpha)}}{\bar{z} - \bar{\alpha}}(-\bar{z}\bar{\alpha}),$$

$$= \lim_{z \to \alpha} \left(\text{conjugate of } \frac{f(z) - f(\alpha)}{z - \alpha} \right)(-\bar{z}\bar{\alpha}),$$

$$= -\bar{\alpha}^2 \overline{f'(\alpha)}.$$

Hence ϕ is differentiable at $1/\bar{\alpha}$.

The solution is now very similar to the proof of Theorem 12; the triangle Δ is decomposed as shown in Fig. 36, and we use the fact that $z = e^{i\theta}$ to establish that

$$\overline{f(1/\bar{z})} = f(z) \text{ on } L.$$

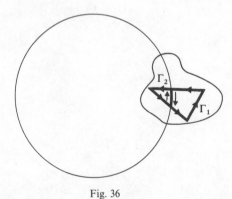

Fig. 36

114

Analytic Continuation by Taylor Series

The Taylor series is a very familiar device for representing function elements, and, if we restrict our attention to series with finite non-zero radii of convergence, then such function elements have discs as their domains. But given one function element in this form how are we to find other function elements?

We shall attempt to answer this question, but first we look at an example.

Example 4

Suppose that we are given the function element (f, D) defined by

$$f(z) = \sum_{n=0}^{\infty} z^n, \quad \text{where } D = \{z : |z| < 1\}. \tag{2}$$

In this case we have a formula $f(z) = 1/(1 - z)$, but in general this may not be so. However, we can use this simple case to illustrate a general idea.

In theory, we can use the series (2) to obtain the value of $f^{(n)}(\alpha)$ for any natural number n and any $\alpha \in D$ (in this simple case it would, of course, be easier to use the formula $f(z) = 1/(1 - z)$.) Thus, for any $\alpha \in D$, we can represent $f(z)$ by the series

$$f(z) = \sum_{n=0}^{\infty} \frac{f^{(n)}(\alpha)}{n!} (z - \alpha)^n, \tag{3}$$

but as yet we cannot be sure where this series converges. For instance, if $\alpha = -\frac{1}{2}$ then $f^{(n)}(-\frac{1}{2}) = (\frac{2}{3})^{n+1} n!$, so that

$$f(z) = \sum_{n=0}^{\infty} (\tfrac{2}{3})^{n+1} (z - \tfrac{1}{2})^n,$$

which is convergent on $D_1 = \{z : |z + \frac{1}{2}| < \frac{3}{2}\}$ (Fig. 37). On the other hand, if $\alpha = \frac{1}{2}$ then $f^{(n)}(\frac{1}{2}) = 2^{n+1} n!$, so that

$$f(z) = \sum_{n=0}^{\infty} 2^{n+1} (z - \tfrac{1}{2})^n,$$

which is convergent on $D_2 = \{z : |z - \frac{1}{2}| < \frac{1}{2}\}$ (Fig. 37).

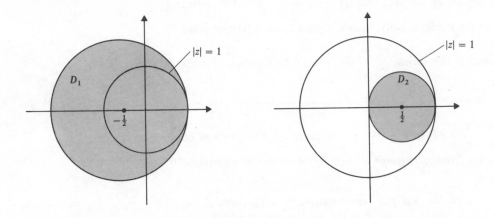

Fig. 37

Thus, for $\alpha = -\frac{1}{2}$, the series (3) converges on a disc D_1 which contains points outside the original disc D, and we have a direct analytic continuation of (f, D); but, for $\alpha = \frac{1}{2}$, we obtain a series convergent on D_2 which is a proper subset of D.

Let us now consider an arbitrary point $\alpha \in D$; then

$$f^{(n)}(\alpha) = \frac{n!}{(1 - \alpha)^{n+1}},$$

115

so that

$$f(z) = \sum_{n=0}^{\infty} \frac{(z - \alpha)^n}{(1 - \alpha)^{n+1}}, \tag{4}$$

which is convergent on the disc $D_\alpha = \{z : |z - \alpha| < |1 - \alpha|\}$ (Fig. 38).

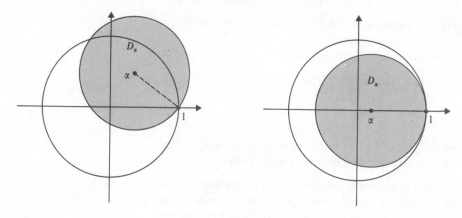

Fig. 38

Notice that for each $\alpha \in D$ the disc D_α has centre α and its boundary passes through 1. If α lies anywhere in D other than on the interval $[0, 1)$ of the real axis, then the series (4) provides us with an analytic continuation of (f, D). In fact, if we choose $\alpha \in D$ on any ray from the origin other than the one which contains $[0, 1)$, then the series (4) provides us with an analytic continuation of f in the direction of that ray. It appears then that the point 1 is acting as a "barrier" to this method of analytic continuation, and this is by no means surprising since $z \longrightarrow 1/(1 - z)$ has a singularity (a simple pole) at 1. We give a precise definition of barrier point as follows.

Definition

> Let the series $f(z) = \sum_{n=0}^{\infty} a_n(z - z_0)^n$ have radius of convergence r, and let α lie within the circle of convergence $|z - z_0| = r$; then the point $z_0 + r\left(\dfrac{\alpha - z_0}{|\alpha - z_0|}\right)$ on the circle of convergence is said to be a **barrier point** of the original series if the series $\sum_{n=0}^{\infty} \dfrac{f^{(n)}(\alpha)}{n!}(z - \alpha)^n$ has radius of convergence $r - |\alpha - z_0|$.

The point $z_0 + r\left(\dfrac{\alpha - z_0}{|\alpha - z_0|}\right)$—$P$ in Fig. 39—is the point on the circle of convergence and on the ray from z_0 through α. (See Self-Assessment Question 10 on page 118.

The barrier points prevent the analytic continuation of a function element defined by a Taylor series, but in a sense it is the barrier points which ensure that the series had finite radius of convergence at the outset.

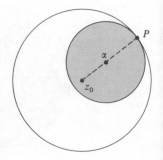

Fig. 39

Theorem 13

Every Taylor series $f(z) = \sum_{n=0}^{\infty} a_n z^n$ with finite non-zero radius of convergence has at least one barrier point.

Proof

Suppose that the series has radius of convergence r, that the inside of the circle of convergence is the disc D, and that the theorem is false. Then (f, D) can be continued analytically beyond the original circle of convergence in the direction of every ray from 0. Thus every point z on the circle of convergence lies in an open disc D_z which is the domain for a direct analytic continuation of (f, D). But the circle of convergence is a compact set and can therefore be covered by a finite number of such open discs. Thus f is analytic on a region which contains the original circle of convergence and D, and this region therefore contains a disc centred at 0 with radius greater than r. But, from Taylor's Theorem, the radius of convergence of the original series must then be greater than r, and so we have a contradiction. ■

It is not a coincidence that in Example 4 the singularity at 1 is also a barrier point, and we ask you to show in Problem 6 below that non-removable singularities on the circle of convergence must be barrier points. There are, however, other kinds of barrier points: for example, branch points.

Example 5

Let $\phi(z) = -\mathrm{Log}(1 - z)$; then $\phi(0) = -\mathrm{Log}\, 1 = 0$, and $\phi^{(n)}(0) = (n - 1)!$. It follows from Taylor's Theorem that $\phi(z) = \sum\limits_{n=1}^{\infty} \dfrac{z^n}{n}$, which is convergent on the disc $|z| < 1$, and clearly the series diverges if $|z| > 1$.

Suppose now that we expand ϕ as a Taylor series about any point $\alpha \in (0, 1)$ so that

$$\phi(z) = \sum_{n=0}^{\infty} \frac{\phi^{(n)}(\alpha)}{n!}(z - \alpha)^n, \tag{5}$$

and suppose that the radius of convergence is greater than $1 - \alpha$. The series (5) defines a function element of $z \longrightarrow -\log(1 - z)$ analytic on a region containing 1, and therefore $\phi'(z) = 1/(1 - z)$ in this region. But we know this to be impossible, and therefore 1 is a barrier point of the original series. In Problem 6 we ask you to show in general that a branch point on the circle of convergence is a barrier point.

The following theorem enables us to use a property of the coefficients of a Taylor series to determine one of its barrier points.

Theorem 14

Let the series $f(z) = \sum\limits_{n=0}^{\infty} a_n z^n$ have unit radius of convergence. If $a_n \geqslant 0$ for all n, then the series has a barrier point at 1.

Proof

There is at least one barrier point, say $e^{i\alpha}$, on the unit circle. The Taylor series for f at $\rho e^{i\alpha}$, where $0 < \rho < 1$, is

$$\sum_{n=0}^{\infty} \frac{f^{(n)}(\rho e^{i\alpha})}{n!}(z - \rho e^{i\alpha})^n,$$

and since $e^{i\alpha}$ is a barrier point, this series has radius of convergence $1 - \rho$. But

$$|f^{(k)}(\rho e^{i\alpha})| = \left| \sum_{n=k}^{\infty} n(n - 1) \cdots (n - k + 1) a_n \rho^{n-k} e^{(n-k)i\alpha} \right|$$

$$\leqslant \sum_{n=k}^{\infty} n(n - 1) \cdots (n - k + 1) a_n \rho^{n-k}$$

$$= f^{(k)}(\rho).$$

Hence, the radius of convergence of $\sum_{k=0}^{\infty} \frac{f^{(k)}(\rho)}{k!}(z - \rho)^k$ is equal to $1 - \rho$, and hence 1 is a barrier point. ∎

Corollary

If $f(z) = \sum_{n=0}^{\infty} a_n e^{in\alpha} z^n$ has unit radius of convergence, $a_n \geqslant 0$ for all n, and α is real, then the series has a barrier point at $e^{-i\alpha}$.

Proof

Let $w = e^{i\alpha} z$ and apply Theorem 14 to the series $\sum_{n=0}^{\infty} a_n w^n$. ∎

Self-Assessment Questions

8. If $f(z) = \sum_{n=0}^{\infty} a_n(z - z_0)^n$ has radius of convergence r, and if $|\alpha - z_0| < r$,

   show that the series $\sum_{n=0}^{\infty} \frac{f^{(n)}(\alpha)}{n!}(z - \alpha)^n$ cannot have radius of convergence less than $r - |\alpha - z_0|$.

9. Is it possible for the series $\sum_{n=1}^{\infty} a_n z^n$ to converge at a barrier point?

10. Show that $z_0 + r\left(\dfrac{\alpha - z_0}{|\alpha - z_0|}\right)$ is on the circle of convergence of $\sum_{n=0}^{\infty} a_n(z - z_0)^n$, which has radius of convergence r.

Solutions

8. The function f is analytic on the disc $\{z : |z - z_0| < r\}$ which contains the disc $\{z : |z - \alpha| < r - |\alpha - z_0|\}$, and it follows from Taylor's Theorem that the series $\sum_{n=0}^{\infty} \dfrac{f^{(n)}(\alpha)(z - \alpha)^n}{n!}$ converges on the latter disc. Hence, the radius of convergence cannot be less than $r - |\alpha - z_0|$.

9. It is possible for the series $\sum_{n=1}^{\infty} a_n z^n$ to converge at a barrier point; for example, if $a_n = \dfrac{1}{n^2}$ then $|a_n z^n| \leqslant \dfrac{1}{n^2}$ if $|z| \leqslant 1$, and $\sum_{n=1}^{\infty} \dfrac{1}{n^2}$ is convergent. Hence $\sum_{n=0}^{\infty} \dfrac{z^n}{n^2}$ converges at every point of the circle $|z| = 1$, and, in particular, it converges at the barrier point (or points) on this circle.

10. It is sufficient to show that

$$\left| z_0 + r\left(\frac{\alpha - z_0}{|\alpha - z_0|}\right) - z_0 \right| = r.$$

The result is immediate, since $\dfrac{\alpha - z_0}{|\alpha - z_0|}$ has unit modulus.

Problems

6. (a) Show that a non-removable singularity or a branch point on the circle of convergence of a Taylor series must be a barrier point.

    (b) If $\dfrac{\text{Log}(2 - z)}{(z - 1)(z + 2i)} = \sum_{n=0}^{\infty} a_n z^n$, what is the radius of convergence of the series?

7. If $f(z) = \sum_{n=0}^{\infty} z^{2^n}$ show that $f(z) = z + f(z^2)$, and show that for any natural number k every point β on the unit circle such that $\beta^{2^k} = 1$ is a barrier point of the series.

Solutions

6. (a) Let $f(z) = \sum_{n=0}^{\infty} a_n(z - z_0)^n$, and suppose that z_1 on the circle of convergence is not a barrier point. If α is a point on the line segment $[z_0, z_1]$ but excluding z_0, then the series

    $$\sum_{n=0}^{\infty} \frac{f^{(n)}(\alpha)}{n!}(z - \alpha)^n$$

    is convergent on a region D_1 containing z_1, and it therefore provides a direct analytic continuation (f, D_1).

    (i) If z_1 is a non-removable singularity then (f, D) is certainly not differentiable at z_1, which provides a contradiction.

    (ii) Any complete chain of function elements in a sufficiently small punctured neighbourhood of z_1 must consist of elements which are direct analytic continuations of (f, D_1). Therefore, by the corollary to the Monodromy Theorem, z_1 cannot be a branch point.

    (b) The function $z \longrightarrow \dfrac{\text{Log}(2 - z)}{(z - 1)(z + 2i)}$ has a branch point at 2 and poles at 1 and $-2i$ (Fig. 40). Of these points, 1 is closest to 0, and therefore the radius of convergence is 1.

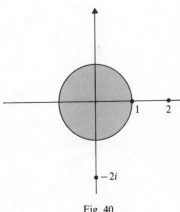

Fig. 40

7. If $f(z) = \sum_{n=0}^{\infty} z^{2^n}$ then

    $$f(z^2) = \sum_{n=0}^{\infty} z^{2^{n+1}} = -z + f(z).$$

    We have

    $$f(z) = z + z^2 + z^4 + \cdots + z^{2^{k-1}} + \sum_{n=k}^{\infty} z^{2^n},$$

    and the first k terms form a finite Taylor series, so that we need consider only the infinite sum on the right. Now notice that $\beta^{2^k} = 1$ so that $\beta^{-2^k} = 1$, and $\beta^{-2^n} = 1$ if $n \geq k$. Hence,

    $$\sum_{n=k}^{\infty} z^{2^n} = \sum_{n=k}^{\infty} \beta^{-2^n} \cdot z^{2^n},$$

    and from the corollary to Theorem 14 it follows that β is a barrier point of the original series. (The roots of $z^n = 1$ are at the vertices of a regular n-sided polygon. Hence, f has barrier points at the vertices of a regular 2^k-sided polygon for all $k \in \mathbf{N}$. If $z_0 \in \{z : |z| = 1\}$, in any open disc centre z_0, however small the radius, there will be barrier points of f, so z_0 itself must be a barrier point. Thus, every point of the circle of convergence is a barrier point, and we say that the circle is a *natural boundary* (or *natural barrier*).)

Instructions for the Construction of the Riemann Surface for log

Cut each of two square pieces of paper from a point midway along one side to the middle of the square. Now place one on top of the other with the flaps over-lapping and glue as shown in Fig. 41. Add as many more pieces as you wish.

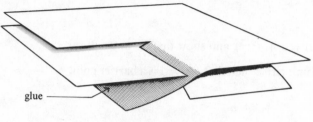

glue

Fig. 41

NOTATION

The following items of notation are explained on the pages given. Most other items of notation used in the text may be found in one of the following: *Mathematical Handbook* for M100, The Mathematics Foundation Course; *Handbook* for M231, Analysis: *Units 1, 2, 3, 4, 5, 6, 8 and 9* of M332, Complex Analysis.

$\tilde{F}$ 25

$\mathcal{L}[F]$ 25

$\displaystyle\int_{-\infty}^{\infty} f(t)\,dt$ 27

$\displaystyle\mathrm{PV}\int_{-\infty}^{\infty} f(t)\,dt$ 27

$\displaystyle\int_{a}^{b} f(t)\,dt$, where f is not continuous at a point in (a, b) 38

$\{f_n\}$ 63

$f = \displaystyle\sum_{n=1}^{\infty} f_n$ 68

$\zeta(z)$ 71

$[h]$ 78

$\Gamma(z)$ 81

(f, R) 82

$\displaystyle\int_{\Gamma} \frac{dw}{w} = \log b - \log a$ 92

$f(z)|_{\Gamma}$ 92

$\log$ 93

z^{α} 96

121

INDEX

COMPLEX ANALYSIS

1 Complex Numbers

2 Continuous Functions

3 Differentiation

4 Integration

5 Cauchy's Theorem I

6 Taylor Series

7 NO TEXT

8 Singularities

9 Cauchy's Theorem II

10 The Calculus of Residues

11 Analytic Functions

12 NO TEXT

13 Properties of Analytic Functions

14 Laplace Transforms

15 Number Theory

16 Boundary Value Problems

Course Team

Chairman:	Dr. G. A. Read	Senior Lecturer
Authors:	Dr. P. D. Bacsich	Lecturer
	Dr. M. Crampin	Lecturer
	Mr. N. W. Gowar	Senior Lecturer
	Dr. R. J. Wilson	Lecturer
Editor:	Mr. R. J. Knight	Lecturer
B.B.C.:	Mr. D. Saunders	

With assistance from:

Mr. R. Clamp	B.B.C.
Mr. D. Goldrei	Course Assistant
Mr. H. Hoggan	B.B.C.
Mr. T. Lister	Staff Tutor
Mr. R. J. Margolis	Staff Tutor
Dr. A. R. Meetham	Staff Tutor
Mr. J. E. Phythian	Staff Tutor
Mr. J. Richmond	B.B.C.
Mrs. P. M. Shepheard Rogers	Course Assistant
Dr. C. A. Rowley	Course Assistant
Mr. D. Sargent	Course Assistant